From the Feed Trough

Essays and Insights on Livestock Nutrition in a Complex World

Woody Lane Ph.D.

ISBN 10: 0-98-332380-1

ISBN 13: 978-0-983-32380-8

Library of Congress Control Number: 2012908610

Lane Livestock Services

240 Crystal Springs Lane, Roseburg, Oregon 97471

www.woodylane.com

To Jeri Frank, my wife and wonderful friend, who proofread all these articles before they left my desk and provided valuable insights and suggestions. For her quiet support, her belief in our life together, her patience, her joy, and her love.

To Doug Hogue, a Cornell professor who saw the potential in a determined young man who walked into his Morrison Hall office on a sunny August Friday afternoon so long ago. To me, Cornell was the shining light on the hill. Doug was a brilliant professor of nutrition and a specialist in sheep production. Doug changed my life. He opened doors, provided guidance, and imparted a sense of enthusiasm and joy of knowledge that has never left me. Without him, I may never have started down this road.

Come to the edge.

We might fall.

Come to the edge.

It's too high!

COME TO THE EDGE!

So they came,

And he pushed,

And they flew.

by Christopher Logue from New Numbers

Jonathan Cape Ltd., London, 1969

Contents

Section 2: Minerals

Section 3: Vitamins

Section 4: Requirements, Ration-Balancing, Foods

Section 5: Feed Tests, Feed Reports, Feed Tags

Section 6: Sheep Production Cycle Nutrition

Section 7: Practical Field Situations

Section 8: Intriguing Research

Section 9: Reflections

Preface

From The Feed Trough... is the name of my monthly nutrition column in *The Shepherd* magazine. I began writing this column in November 1993. Guy and Pat Flora, the publishers of this fine magazine through 2011, and Cat Urbigkit, the current publisher, have been incredibly supportive and patient, because my column does not always stay within the strict confines of nutrition, and it often arrives on their desk barely in time for publication.

At twelve articles each year, my portfolio grew steadily to more than 200 columns. Over the years I've used many of these articles as handouts in my courses and workshops. Feedback letters are not very common, but in a recent course, one rancher student said to me, "Hey, I finally read those articles, and they're pretty good!" I'll take complements as they come.

After years of monthly columns, I finally decided to compile many of the nutrition articles in a book — this book. We all enjoy magazine articles, but we also know what happens to magazines over time. In spite of our best intentions, magazines lie on shelves and gather dust for a few years. Finally someone in the house exclaims, "Why on earth are we keeping these?" and throws them out. Or a determined reader carefully snips articles and puts them into a file labeled "Miscellaneous." Ten years later that file sits in a forgotten box in the garage. But by compiling these columns into a book, I've tried to put them on solid ground. More folks can enjoy these columns in a book, learn from them, consider their points of view, and locate them again years later on a bookshelf.

I've tried to make this book useful and also user-friendly. In addition to a standard *Table of Contents* which lists titles and page numbers — you'd expect that, wouldn't you? — I've also included a one-phrase description of each article's topic, so you can skim down the titles and see what each article is about. There's also an index. And finally, each article contains a brief *Author's Note* in which I add something relevant — background, color, update, or perhaps a bit of soapbox.

Each article is an independent document. In some cases, articles written years apart may focus on related topics. Which means that you may encounter some redundancies — like the acronym NDF defined in a couple of different articles. My hope is that the definitions are the same. Also, a couple articles are on topics that now may be historical, although they were current when I originally wrote about them, such as a legal controversy concerning selenium or the limitations of a reference book published in the 1980s. Have patience. These articles are still interesting and informative.

One thing you should note: *This book is not a primary textbook on animal nutrition or even on ruminant nutrition.* Primary textbooks are those hoary college tomes in which principles and facts are methodically organized across hundreds of pages. Primary textbooks are designed for college students who must memorize everything and then cram for exams. No, my book is a *supplementary text* — back stories, unusual perspectives, bite-sized articles that make you think. I sincerely hope that this book is something you *want* to read, in contrast to all those academic textbooks you were *forced* to read in school.

This book is also a general book about livestock nutrition; the topics apply to all species. True, there are a lot of articles that relate to sheep — you would expect that in articles written for a sheep magazine — but anyone interested in livestock of any species will gain from these topics. NDF is NDF no matter who consumes it, and my articles on topics such as TDN or minerals or hay or cold stress apply to *all* livestock species.

In any case, this is *your* book. Read it in any style you want. Consider how people eat corn-on-the-cob. Some folks methodically eat from one end to the other; some eat round and round; and some eat, well, randomly, nibbling here and there, whatever strikes their fancy. Like a bucket of hot buttered corn-on-the-cob at a summer picnic, this book is for savoring and enjoying in any way you like.

I've written these articles to inform. I've written them to convey good information in a readable and understandable way, and also, I hope, to entertain. I believe that people remember things better when they enjoy them. So I've generally used active voice and common words. And declarative sentences — subject, verb, object — subject, verb, object. Except for those times I haven't. And occasionally there may be some humor, at least I think so.

But most of all, these articles are my way of sharing good information about important nutritional topics. I firmly believe that knowledge gives us control, and that we make good decisions when we know as much as possible. Nutrition is a complex topic. Knowing about nutritional principles gives us the tools to deal with the practical situations we encounter in the real world of farming and ranching.

So with great respect for nutrition, I've written these 89 short essays. Sometimes they convey standard perspectives, sometimes they don't, and sometimes their perceptions may seem — how shall I say it — a bit unusual. Well, I want you to think about these topics, to contemplate their various turns and twists, to think outside of the box, and if you will, to ruminate.

There's lots of food at this feed trough . . .

Woody Lane
Roseburg, Oregon
April 2013

About the Author

Woody Lane is a nationally known livestock nutritionist and forage specialist living in Roseburg, Oregon. He owns and operates *Lane Livestock Services*, a consulting firm based in the Pacific Northwest.

Woody is one of the most active and independent livestock nutritionists in the country. Among his many activities, he organizes and teaches courses in livestock nutrition and forage management to ranchers in Western Oregon, facilitates forage study groups for farmers and ranchers, and writes the popular monthly column "From The Feed Trough..." for *The Shepherd* magazine. This book is a collection of some of those columns.

Woody earned his Ph.D. and M.S. degrees in Animal Nutrition from Cornell University. He has published hundreds of popular articles and fact sheets on sheep and cattle production and grazing management and more than 25 research articles in peer-reviewed scientific journals.

Originally from New York, Woody earned an undergraduate B.S. degree in Zoology from Syracuse University and served for two years as a Peace Corps Volunteer in Sarawak, Malaysia. After his graduate studies at Cornell in the 1970s, he worked on the famous Allegheny Highlands Project in West Virginia. This project was the innovative prototype for the Integrated Resource Management programs that are used throughout the livestock world today. In the 1980s, he served on the faculty of the University of Wisconsin-Madison as the State Sheep and Beef Cattle Extension Specialist. He and his wife Jeri Frank then moved west and have lived in western Oregon since 1990.

Woody is an expert on sheep and beef cattle nutrition, pasture management, and grazing techniques. An exciting and skilled speaker, he has been featured in many of the top workshops and conferences across the United States and Canada. His clients are based all over the U.S. and across the world.

Among other major projects, Woody helped develop the recent editions of the classic *SID Sheep Production Handbook*. He was also the operations manager for the American National Sheep Improvement Program (NSIP), and together with veterinarian Don Bailey, created the highly acclaimed instructional videotapes *Lambing Time Management.*

When not working with livestock and pastures, Woody enjoys dancing and calling contra dances and square dances. But that, as they say, is a different story altogether.

SECTION 1

Energy, Protein, Fiber, Anatomy

Nutritional Architecture

One of the first things I teach in my Livestock Nutrition class is how to tell the difference between dogs, horses, llamas, and sheep — from the inside out.

Animal nutritionists view livestock differently than everyone else; we don't see breed styles or straight backlines or colored fiber. Instead, we see gastrointestinal tracts (GI tracts) mounted on four legs. Animal nutritionists reduce animals to their most elemental level — you could say that we go directly to the guts of the issue.

Animals boast a wide array of digestive tract designs that are directly related to their diets. Dogs eat meat; horses and sheep eat grass; bears and skunks eat anything that moves and a few things that don't. Goats, incidentally, do *not* eat tin cans. (Tin cans are now made of aluminum, and goats don't have a nutritional requirement for aluminum.)

Let's begin with the dog. A dog's GI tract is a relatively straightforward affair, literally. Once food is consumed, it goes down the gullet (the *esophagus*) into the true stomach (the *abomasum)*. The abomasum contains hydrochloric acid for low pH and the enzyme *pepsin* for the initial digestion of protein. After a short time in the abomasum, ingesta moves into the next section of the GI tract, which is the small intestine.

The small intestine is a long set of thin loops that many people call "guts." The major work of digestion takes place here. The pancreas and other organs secrete most of their digestive enzymes into the small intestine, and a high proportion of nutrient absorption takes place across the intestinal walls. Interestingly, the small intestine is not really small at all. It is quite long, which makes sense because a longer tube contains more surface area, which increases the animal's capacity to absorb nutrients. For example, a dog maintains more than 13 *feet* of small intestines. Even rabbits have 12 feet of them.

After passing through the small intestine, undigested material moves into the large intestine (the *colon)*. This is only a medium-sized organ in dogs and humans, but it can be quite large in other species. In a dog's colon, water is extracted or added to the residual material, which then moves into the rectum for excretion.

A pig is slightly more complex than a dog. While the front end of a pig is about the same as a dog, the hindgut is quite different. After passing through more than 60 feet of small intestines, ingesta enters a colon that is *much* larger than a dog's. The pig's colon contains a large and active population of bacteria and protozoa that ferment (rot) some of the organic matter. The end-products of this process are nutrients that the pig absorbs across the wall of its GI tract. Thus, microbial fermentation in the colon allows the pig to extract some nutrients from an otherwise poorly-digestible material, like fiber.

A horse, from our perspective, is really a large pig that is more efficient at digesting fiber than a pig. A horse contains a *humongous* colon, which is much larger than a pig's. Ingesta can ferment in a horse's colon for days and days. Such a long fermentation period allows a horse to digest fiber far more efficiently than a pig. The horse has evolved well. The horse can enjoy the lifestyle of a grazing animal because its GI tract allows it to successfully obtain enough nutrients from forages.

In essence, the horse's colon is a large fermentation vat, kind of like a flow-through septic tank on legs. In a horse, this compartment is placed at the far downstream end of the GI tract, just before excretion. However, if you were a nutritional architect and had the opportunity to design a GI tract from scratch, you might consider it more efficient to put the fermentation compartment *at the beginning of the pipe, rather than at the end* — so that all the products of microbial fermentation and even the microbes themselves could be subjected to the digestive enzymes of the stomach and small intestine. Well, that's precisely what has happened in a ruminant.

The main fermentation compartment in ruminants is called the *rumen.* Over millions of years of evolution, the ruminant stomach differentiated into *four* compartments: the rumen, the reticulum, the omasum, and the abomasum (the true stomach). Feed travels down the esophagus into the rumen and ferments merrily there for days. The *reticulum* is so tightly attached to the rumen, that these two organs are usually considered a single unit (the *reticulo-rumen*). Ingesta then passes into the *omasum.* While no one really knows what the omasum does, physiologists think that it may have something to do with water balance. In any case, from there ingesta moves into the true stomach, and you know the rest of the story from here . . .

This design is so efficient that it makes the horse's digestive system pale in comparison. Ruminants are masters at extracting nutrients from forages and other feedstuffs. Not only does feed have *two* chances to digest (once in the rumen and once in the lower tract), but nature has even given ruminants a recycling system: *cud.* This is ingesta that ruminants regurgitate up from the rumen for another chew. (You've never seen a sheep ask for a Copenhagen, have you?

They don't have to). By chewing their cud, ruminants can crush and grind feed particles again and again, especially tough fiber, which makes it easier for the rumen microbes to ferment.

The literal meaning of *ruminant* is an animal that chews its cud. Sheep, cattle, deer, goats, and bison are all ruminants. The largest ruminant? The giraffe. Consider, for a moment, a giraffe's engineering design — it must move cud from its rumen all the way up to its mouth.

It makes one pause.

Other species do not chew cud. Dogs are not ruminants, and even the most badly behaved dogs do not chew cud. Cats don't chew cud (hair balls do not qualify as cud). Horses also don't chew cud. Remember that a horse's fermentation compartment is at the *far end* of the GI tract, away from the mouth. Therefore, if you ever see a horse chew its "cud" or whatever, call your veterinarian.

And llamas? Technically, llamas are not true ruminants (suborder *Ruminantia*), although they *do* have a rumen and they *do* chew their cud. Llamas actually belong to an offshoot evolutionary family of *camelids* (suborder *Tylopoda*), which also includes camels, vicunas, and alpacas. But here's the difference: camelids don't have an omasum. Camelid stomachs have evolved into only three compartments, not four.

Therefore, if you ever look at a cud-chewing animal and don't know what it is, just remember that a sheep has an omasum and a llama does not. But just for security, you might want to move out of spitting range.

First Published: October 1996

Author's Note: These two principles are fundamental to ruminant nutrition: the microbial fermentation of substrates in the GI tract, and the places along the GI tract where this fermentation occurs. In an article like this, I may portray these principles in a plain and breezy manner, but one should look past the informality. A person who understands these concepts and appreciates the architectural considerations for digestive tract design is a person who grasps the entirety of the digestion process and has the knowledge to manipulate diets in accordance with the principles of ruminant and non-ruminant nutrition.

Ruminant Ruminations

When I submit a hay sample to a laboratory for nutrient analysis, the report may come back with *two* different values for TDN (*Total Digestible Nutrients*). The higher value is for ruminants like sheep and cattle; the lower value is for horses, and they may differ by more than ten units. TDN is a measure of a feed's energy value. Can one hay sample really have two different energy values? What's going on here?

The short answer is, yes. Hay *can* have two energy values, or even more than two, depending on who is consuming that hay. To explain this we must look at the architectural differences between ruminants and non-ruminants.

Let's think, for a moment, about one of the most compelling principles of life — *food.* Really the search and acquisition of nutrients. Quite simply, food is one of the driving forces of evolution. The need for food controls some of the main design features of animals.

The second critical principle is the *type* of food. Of all the edible materials in the world, what is the most abundant? What potential feed material is so common that it could support millions of animals with enough energy for survival, reproduction, and other evolutionary useful activities? It's not sugar. It's not protein. It's cellulose. *Fiber.* The bulk of most green plants is composed primarily of cellulose and other fibrous compounds. But there is a catch: even with all this potential feed around, no mammal can use it directly — because no mammal makes the *cellulase* enzyme necessary to digest this fiber. Bummer.

But over the millennia, nature devised an elegant solution. Although mammals don't secrete the enzymes for fiber digestion, many microbes do. These microbes can digest fiber by a process called *fermentation.* Therefore, mammals who spend their time eating forage have evolved gastrointestinal tracts containing regions that house these microbes, areas where microbes can cheerfully ferment the fiber. In turn, those mammals can digest the nutritious products of microbial fermentation and even gain nutrients from the microbial bodies themselves. It is a brilliant two-stage system that allows forage-eating mammals to derive their nutrients from fiber. Other species that have prowled the earth — like saber-tooth tigers and cave bears and man — evolved a different type of GI tract. They didn't need to digest much fiber because they simply ate the animals who did.

And that brings us to the problem of TDN for horses.

Put yourself in the position of a gastrointestinal architect: your job is to adapt a GI tract to digest fiber. Your major tool is a sac containing microbes that are good at fermenting things, and you must insert that fermentation sac somewhere along the GI tract. However, you also must obey one immutable rule for all mammals: the true stomach (*abomasum*) must be positioned in front of the small intestine with nothing inserted between them. You cannot alter this arrangement. Therefore, you really only have two main choices — you can either put the fermentation sac *in front* of the stomach or you can put it *after* the small intestine. (Here, "in front" means upstream and "after" means downstream. The technical terms for these choices are *anterior to the abomasum* and *posterior to the small intestine.*)

The first choice gives us the basic architecture of ruminants like sheep and cattle. All their feed *first* goes into their fermentation sac — called a rumen — *before* it enters the true stomach. Ruminants also have the ability to chew cud, which means that they can regurgitate rumen contents for another chew, which causes further physical breakdown of the fiber. Rumen fermentation of this fiber produces metabolic products which are absorbed across the rumen wall. After leaving the rumen, the residual feed mass passes into the true stomach and small intestine for further digestion. Also, the bodies of dead rumen microbes flow out of the rumen into the stomach and small intestine, where they provide even more nutrients for digestion. Ruminants, clearly, are designed to extract a maximum amount of energy from fibrous feeds. From an engineering perspective, a ruminant's digestive tract features multiple opportunities and sequential digestion (rumen and then small intestine) and redundant mechanical systems (chewing cud).

In contrast, the second choice gives us the basic architecture of the horse. Its fermentation sac is the large intestine (called the *colon*), which in horses is *very* large. The basic design of the horse is that feed *first* goes into the true stomach and small intestine *before* it passes into the colon. This means that digestible nutrients, like starch and sugars and most proteins, are digested and absorbed from the small intestine just as it is in humans. Fiber, on the other hand, cannot be fermented until it travels all the way through the GI tract and into the colon. Microbes in the colon ferment this fiber, and then the metabolic products of this microbial fermentation are absorbed across the colon wall. The residual mass then leaves the colon and becomes manure. Unlike ruminants, however, horse architecture provides no second chances for additional digestion of fiber or microbial bodies.

Also, in case you've never noticed, horses don't chew cud. Which means that horses can't smash and grind fiber as well as cattle and sheep. Because horses cannot crush fiber as well as ruminants, they don't ferment fiber as extensively as ruminants. In other words, fiber is usually less digestible in horses than in ruminants. This characteristic becomes increasingly important as the feed contains more fiber or contains fiber of poorer quality.

There we have it. The reason for the two energy values in a feed is the basic architecture of the gut. Sheep and cattle are more efficient than horses at extracting energy from fiber. TDN is essentially a score of the ability to extract energy from feeds. The differences in TDN between ruminants and horses reflects their differing efficiencies of fiber digestion. This TDN divergence becomes greater as the percentage of potentially-digestible fiber in feeds increases. For example, a good quality pasture (i.e., young grass at 20% protein with low fiber levels) can have TDN values of 65% for ruminants and 61% for horses — a difference of only four units. But a poorer quality hay (i.e., first-cutting grass hay at 10% protein with high fiber levels) can have TDN values of 58% for ruminants and only 45% for horses — a difference of thirteen units.

Good reference tables and laboratory reports list separate TDN values for the various species. But some publications still contain only one TDN value per feedstuff. So if you're trying to balance diets for sheep and horses, and you only have one TDN value, how do you apply this value to both species? Maybe you should find a better reference table.

First Published: November 2000

Author's Note: Fascinating, just fascinating. The rumen is a large, bulky fermentation sac that ruminants fill during long continuous grazing sessions and then spend many hours chewing their cud. An abomasum is not as elastic as a rumen, so horses cannot eat as much at one time as cattle, so horses nibble more often during the day. A horse's colon is large but not as bulky as a rumen. The difference comes out in the run. Horses can run rather well, and cattle run like, well, cattle. Have you ever seen a cow run in the Kentucky Derby?

Second Time Around

As ruminants and horses graze and browse their way through the plant world, they depend on fiber fermentation as their main source of nutritional energy. Their digestive tracts are marvels of specialized adaptation. The rumen in sheep and cattle and the large intestine (colon) in horses contain huge populations of microbes which ferment fiber into nutrients that are absorbed by their hosts. But ruminants and horses represent only *two* design strategies used by herbivores to extract nutritional value from forages. Nature has actually devised other fiber fermentation strategies. Here's one of them . . .

To ferment fiber successfully, animals must house their microbes in a fermentation sac. Ruminants depend on a rumen which is located *in front* of their true stomach. Horses depend on their large intestine, which is located just *after* their small intestine. But let's think back to our high school biology class. Remember that we humans have a tiny blind-sac appendage at the *junction* of our small intestine and large intestine? This annoying, residual organ is our *appendix.* It is residual because humans have no functional use for it and annoying because it occasionally becomes infected and develops into appendicitis.

The official anatomical name for this gastrointestinal blind alley is the *cecum.* Many animals have evolved to use the cecum as their fermentation sac. In those species, the cecum is much larger than it is in humans. After digestive material leaves the small intestine, some of it flows into the cecum where it ferments — in a process similar to that occurring in a rumen. Microbes flourish in the cecum, and their nutritive fermentation products are absorbed across the cecal wall. After a short retention period in the cecum, the residual digesta flows out of the cecum through that single opening, together with some microbial debris, and then it passes through the large intestine for excretion as manure.

So, which animal species depend on cecal fermentation? Well, the cecal fermenters include two of the most common orders of mammals: rodents and rabbits. Nature wasn't stingy with rodents. There are more than 1,700 species of them: mice and rats of course, and porcupines, muskrats, beaver, gophers, guinea pigs, gerbils, squirrels, lemmings, prairie dogs, and lots of others. Although rodents generally eat nuts and seeds, they also have the ability to digest some fiber. We've all observed with dismay what mice will do to a stack of old newspapers. And the larger South American rodents — the capybara and nutria — are true grazers who consume forages just like sheep and horses. The semiaquatic capybaras, by the way, are not small. Adults can be four feet long and weigh 100 pounds.

The other group of cecal fermenters is the rabbits, but we'll come back to them in a moment.

Here's a question: if a significant portion of the world's animals successfully use cecal fermentation as their nutrition strategy, then why don't our standard livestock species use it instead of relying on rumens or large intestines? There are two main reasons. The first is that the cecum is a blind sac; *all solid material must enter and exit the cecum through the same opening.* Anyone who travels by airplane is all-too-familiar with this scenario — you're trapped in that metal tube with only one exit until all the passengers in front of you gather their bags and walk out, regardless of how much you fret about your tight connection. In rodents, like in the airplane, only a small amount of digesta can easily pass through the cecal opening. This bottleneck limits the amount of fiber that can be fermented. Ruminants and horses, in contrast, have wide flow-through fermentation organs that allow large amounts of digestive material to enter and exit easily.

The second reason is the cecum's actual location in the GI tract. This concept becomes clear if we think of the gastrointestinal tract as a long continuous tube that begins at the mouth and ends at the rectum. The main absorption region of this tube is the small intestine, where most of the gastrointestinal enzymes for digesting carbohydrates and proteins are secreted. The cecum is located *after* the small intestine, downstream from the region containing all those enzymes. This location has a couple of notorious limitations: (1) most of the easily fermentable carbohydrates (like simple sugars) never reach the cecum because they are absorbed upstream in the small intestine, and (2) all of the potentially valuable protein and energy contained in the cecal microbial debris is lost in the manure.

Therefore, cecal fermenters suffer from the combination of a blind-sac bottleneck and the inherent inefficiency of the cecum's location. The bottom line is that cecal fermenters can obtain *some* nutritional energy from fiber, but not as much as ruminants or even horses.

But one group of animals — the rabbits — evolved an elegant (some would say *inelegant*) strategy for improving their fermentation efficiency. They *recycle* some of their manure and run it through the digestive system for a second time.

How can I describe this delicately? Well, I can't, so I'll be straightforward about it. We're talking, of course, about manure.

Rabbits excrete two different types of manure: a hard pellet and a soft pellet. The hard pellet contains indigestible fiber and waste products. This is the true waste material that drops to the ground and remains there, just like sheep pellets or cattle manure. But the soft pellet (night feces) has a different purpose. The soft pellet contains the residual contents of the cecum and other material of nutritional value, such as potentially-digestible fiber, microbial debris, B-vitamins, etc. The nutritional goal is to expose this material to the digestive enzymes of the small intestine. Rabbits achieve this goal by *consuming* the soft pellet as it is excreted, thus giving the material a *second* chance to pass through the digestive tract. When that material reaches the small intestine, it is digested and absorbed, thus supplying the rabbit with energy, protein, and vitamins. The process of consuming fecal material is called *coprophagy*, and it's actually a sound evolutionary adaptation to the inherent inefficiencies of cecal fermentation. Some would argue too sound. Rabbits are very good at consuming forage. Ask any Australian.

But let's reflect for a moment on the types of animals that depend primarily on cecal fermentation for fiber digestion, including the species that practice coprophagy. None are very large. Most of the smaller species, including the rabbits, also consume nuts and grains and other low-fiber material in their diet. While cecal fermentation is indeed one strategy for deriving nutrition from forages, it is not efficient enough to support the larger herbivores. Those animals have placed their microbes in the rumen or the large intestine.

So now we've discussed *cecal fermenters* like rodents and rabbits, *rumen fermenters* like sheep, cattle, and giraffes, and *large intestine fermenters* like horses, rhinos, and elephants. But what about the hippopotamus? Hippos live on forage, and they certainly don't chew cud or practice coprophagy. In fact, hippos don't fit any of the three main fermentation categories. But let's save hippos for another time, because a hippo is, uh, a horse of a different color.

First Published: December 2000

Author's Note: When you really think about it, this all makes perfect sense. But nonetheless, I'm very glad that elephants never discovered coprophagy.

A Calorie By Any Other Name . . .

Go to your local supermarket and read the "Nutrition Facts" label on the side of a box of crackers. What is the energy value of those crackers? I recently saw a box that listed 140 Calories for a serving of 31 grams, which equates to 4.5 Calories/gram. How valuable is this number? The cardboard box itself has an energy value of 4 Calories/gram — nearly the same caloric value as the crackers in it. So which is a better feedstuff — the crackers or the box? Since they are nearly equal in calories, which one should we use with our onion dip?

Actually, this isn't a joke. That cracker box label illustrates a very basic nutritional problem — how can we express the energy value of a feedstuff in a nutritionally meaningful way, so that we can balance diets reliably? We know intuitively that crackers are more nutritious than their boxes, but what number can we use that will properly reflect this?

Nutritionists recognized this problem more than a hundred years ago, and they derived some good answers that we still use today. Although they didn't have our modern computerized tools (they were still waiting for Windows 1888), they could accurately measure energy with a piece of equipment called a *bomb calorimeter*. This is basically a steel cylinder immersed in a container of water. Scientists carefully weighed a small feed sample, placed it at the bottom of the cylinder, and burned it with an electronic ignition (the bomb). All the sample's organic matter would burn completely in seconds, leaving only a small pile of ash. The resulting heat caused a slight increase in the temperature of the surrounding water, and scientists then used basic physics to convert that rise of temperature into calories.

But feedstuffs don't just have one nutritional energy value; they have four: *Gross*, *Digestible*, *Metabolizable*, and *Net*. So here we go . . .

When we measure the energy of a feedstuff *before* it is consumed, with no correction for its digestibility or other biological losses inside the animal, that energy value is called *Gross Energy* (GE). This is the total amount of calories in the feed, period. *This* is the number on the labels in our supermarkets. Crackers and cracker boxes give approximately the same number of GE calories.

But really, this GE value is not very useful because it tells us nothing about the biological *availability* of those calories. When samples are burned in a bomb calorimeter, all carbohydrates essentially give the *same* GE value — 4 Calories/gram. All proteins give slightly more than 5 Calories/gram. And all fats (that are in the form of triglycerides) give 9 Calories/gram. Thus, GE doesn't differentiate between the nutritional values of corn starch, its cardboard box, or an old bamboo fishing pole.

In the late 1800s, livestock nutritionists refined GE by looking at the other end of the animal. They reasoned that any calories remaining in the manure had passed through the intestinal tract without being absorbed. Clever fellows. Therefore, by subtracting the energy value of the feces from the energy consumed, they calculated the amount of energy that was absorbed — digested — across the intestinal tract. (Imagine, for a moment, what it meant for someone to perform bomb calorimetries on those thousands of fecal samples.) Nutritionists called this value *Digestible Energy* (DE), which they carefully defined as the Gross Energy minus the energy lost in the manure.

Today, DE is still one of the most useful values in livestock nutrition. DE often goes by different names, which are all more or less interchangeable, although sometimes we have to convert units or make minor corrections. DE is sometimes called Digestible Organic Matter (DOM), Digestible Dry Matter (DDM), and our venerable favorite, Total Digestible Nutrients (TDN). For those people who enjoy playing with spreadsheets, you can convert DE to TDN with the rule of thumb that 2,000 kilocalories (2 Mcal) of DE equals approximately one pound of TDN.

But we cannot stop here. DE is *not* the final amount of useful energy in a feed. Calories are lost in more ways than in manure. Two other obvious routes are gas and urine.

Gas? Oh yes, all livestock lose feed energy in gases. Ruminants make a livelihood of it because of the fermentation in their rumen. Two major products of rumen fermentation are methane and carbon dioxide. (Don't smoke in the rumen of a sheep.) But non-ruminants also lose energy in gas, although on a much smaller scale. If you don't believe me just, uh, eat some beans.

The second loss, urine, contains energy in the soluble compounds flushed out during excretion. The main soluble compound for mammals is urea. Although these calories may have been cycled through the liver and kidneys enroute into the urine, once they are captured in urea, they are not used for muscle, heat, or milk. Therefore they represent a loss to the animal.

These urine and gas losses give us another term: *Metabolizable Energy* (ME). More precisely, ME is the energy value of the feed after subtracting the energy lost in the manure, gas, and urine. ME is *always* smaller than DE which is *always* smaller than GE — because each term reflects subtracting more calories that are lost to the animal.

But we're not done yet. Livestock still cannot use all the calories in ME — at least not mammals or birds. Because these are warm-blooded species, they produce heat that requires energy. But energy used for heat cannot be used to create milk, wool, or muscle. So nutritionists defined a fourth major category of feed energy: *Net Energy* (NE). Finally, after accounting for all the losses in manure, gas, urine, and heat, *these* NE calories are the calories available for tissue production — i.e., the *net* amount of energy in a feed. Of course, heat losses are very difficult to measure accurately. Animals must be put into complex metabolism chambers and monitored . . . well, that's a story for a different time.

Over the past 30 years, NE has spawned a whole family of related terms based on the way it is used. Some NE calories are used for creating milk and wool and tissue, and some NE calories are just reserved for maintaining the animal (for basic functions like breathing and walking and eating, etc.). Therefore, nutritionists coined the terms *Net Energy for Maintenance* (NE_m), *Net Energy for Gain* (NE_g), *Net Energy for Lactation* (NE_l), and a few others that you'll find in the literature. These terms allow nutritionists to allocate precise amounts of energy to different types of production. Good for balancing rations and for graduate student research projects.

I've introduced a lot of acronyms — GE, DE, DDM, DOM, TDN, ME, and the whole family of NE values. But here's something to ruminate on: In supermarkets, every box of human food lists only GE on its label. It's the law. Yet no one in livestock nutrition uses GE — animal scientists climbed out of that box a long time ago.

First Published: November 1998

Author's Note: We are only scratching the energy surface. Every nutrition student learns these energy terms in his first course in animal nutrition. The true flux of calories and the efficiencies with which they are used are the intricate topics of thousands of dissertations and professional papers. But at least after this article, you can begin to have an inkling that providing energy to animals (and humans) is not as simple as it may first appear.

TDN

If you're going to buy a feedstuff, you need to know its energy value. But energy values are expressed in all sorts of cryptic terms: TDN, DE, ME, NE_g, NE_m. This can be worse than a stockbroker's report. Let's look carefully at one term that is used nearly everywhere . . .

The granddaddy expression of energy terms is the well-known and venerable "TDN" — *Total Digestible Nutrients.* As the name implies, TDN was designed to express the total amount of available nutrients in a feedstuff. TDN is based on the concept of "digestibility," which simply means the proportion of a feedstuff absorbed as nutrients across the gut wall after being broken down in the intestinal tract by enzymes and microbes. The undigested portion of the feedstuff continues down the gut towards the exit.

The TDN system did not spring full-blown from an ancient Sanskrit text. German and American scientists originally devised the TDN concept around the turn of the last century. Agricultural scientists had been searching for a single number to reflect the nutritional value of a feed. So, based on the limited chemical knowledge at that time, they decided to partition feed organic matter into four nutritional fractions: *crude protein, crude fiber, crude fat,* and *nitrogen-free extract (NFE),* which was defined to mean everything not accounted for by the other three categories. They calculated a total digestibility (TDN) by summing the digestible amounts of each fraction, although they first multiplied the fat fraction by 2.25 because fat contains 2.25 times more digestible energy than carbohydrates or protein. Nutritional reference tables typically list TDN values as percentages or pounds. The basic TDN formula has not changed in over eighty years.

Today, many researchers scorn the TDN system because several of its assumptions are suspect or routinely violated. This skepticism has some merit. For example, NFE is calculated by subtracting the values for water, ash, protein, fat, and crude fiber from 100. This virtually guarantees that any errors in the other values are automatically passed along to the NFE value. It's somewhat like an old-fashioned bucket brigade with the person at the end getting stuck holding the final bucket.

Another problem is the concept of NFE, which was originally *supposed* to represent the more digestible carbohydrate fraction of a feed, which are primarily starch and soluble compounds. The problem is that, in reality, the NFE fraction also contains lignin, which has a digestibility of zero. Conversely, the crude fiber value of a forage can be very misleading because the procedure for analyzing crude fiber *does not detect* certain types of indigestible fiber, like lignin. Modern analytical techniques are much better at identifying a feed's nutritionally important fiber — like neutral detergent fiber (NDF) and acid detergent fiber (ADF), but these new fiber values are not included in the TDN equation. In fact, the "fiber" number legally required on most feed tags is still crude fiber.

Because forages contain many different types of fiber, the TDN system tends to underestimate the true digestibility of high quality forages and overestimate the true digestibility of low quality straws.

TDN also assumes that all fat contains 9 kcal/g of energy. Unfortunately, this assumption is true only for fats like triglycerides, which are the major storage compounds in feed grains. Forages and byproduct feeds can contain all sorts of other fatty substances, such as steroids and waxes, that contain less energy than triglycerides. In the pasture world, however, this inaccuracy doesn't amount to much because forage diets usually contain very little fat.

TDN also includes an imprecision with protein, although this fault is self-correcting. The TDN formula assumes that crude protein contains the same energy level as carbohydrates (i.e., 4 kcal/g). Protein actually contains more energy than 4 kcal/g, but some of that energy is lost into the urine *after* digestion. By using the lower number of 4 kcal/g, the TDN formula cleverly includes a 30% correction for this loss. This makes the TDN value somewhat more accurate than its reputation. Many researchers usually miss this interesting point.

Finally, TDN values are calculated from measurements on individually-confined animals fed a maintenance diet. This means that feed consumption is restricted and constant, and that the animals are not growing, lactating, or pregnant. But how many ranches run herds or flocks of confined animals on maintenance diets?

Actually, the real questions become: how do "real life" feeding situations on productive farms and ranches affect these TDN values? And how do we compensate for these changes?

We know that as feed intake rises above maintenance levels (i.e., for every sheep and cow that grows, lactates, or reproduces), the rate of passage of this feed also increases, which reduces the time fiber spends in the rumen, which reduces its fermentation time, which reduces its digestibility. On the other hand, increased feed intake also increases the total amount of nutrients available in the gut. On the other hand, if given a chance, sheep and cattle will always "sort" their feed, which is a fancy way of saying that they'll eat the cream every time. As scientists are inclined to say, "Life is simpler at maintenance."

Even back in 1920, everyone knew that TDN wasn't perfect. Scientists often pointed out its shortcomings and over the years developed some alternatives. Yet with all its blemishes, the TDN system is still around today. Ranchers and extension agents all across North America successfully use the TDN system to balance diets and purchase feeds. Why?

Because TDN values are simple and practical, and for most animals most of the time, they work. There are more sophisticated energy systems for dairy farms and feedlot calves and lambs, where feed composition and intake are carefully measured and controlled. However, for many sheep and beef cattle operations, where ranchers typically feed with a front-end loader or unroll big bales down a hillside, TDN's imperfections have minimal impact.

Also, many standard nutritional reference tables typically list energy requirements in TDN units. The process of balancing a diet *requires* that the animal's nutrient requirements and the nutritional qualities of the feedstuffs are all expressed in the *same units*. In most cases, nutrient requirements have only been independently determined in TDN units. But those are reasons why TDN works — some inaccuracies have a way of compensating for each other, and there are also a whole lot of TDN values that were originally based on good observations.

So in the practical world, TDN is really not a bad choice after all. That is, unless you have to work with a front-end loader calibrated in decimal places or a nutritional reference table that lists energy requirements in terms of light years per kilogram.

First Published: December 1994

Author's Note: The energy world is confusing indeed, and exceedingly complex. But in the end, when you are sitting freezing on a tractor on a cold winter day, trying to move a big bale across a hillside, the simplicity and understandability of TDN is very attractive.

Ruminations on Protein

Ruminate — Ru-mi-nate — 1. to chew the cud. 2. to meditate or muse, to ponder. (The American College Dictionary, 1970).

So we're going to ruminate a little here . . . I intend to fill this article with all sorts of interesting bits of information, some practical, some just interesting, mostly about nutrition. Like a feed trough. We'll ruminate a bit, chew the cud, and maybe come up with something a little different. Anyone who has taken my rancher courses here in western Oregon knows that nutritional things aren't always what they seem. So let's get started.

First, we'll talk about protein . . .

Look at any feed tag, especially at the "Guaranteed Analysis" section of the tag. Find the term *Crude Protein.* Do you know what it *really* means? More importantly, do you know what it *doesn't* mean?

Glad you asked.

Well, feed protein isn't called "crude" because it spits in public and doesn't attend the opera. Crude protein is called "crude" because it's a crude approximation of the real protein value. To put it another way, when a feedstuff is analyzed for "protein," it isn't analyzed for real protein; it's actually analyzed for something else which is related to protein. Who said that life should be simple?

The dilemma is that you can't easily or accurately measure true "protein," at least not at the speed and scale required by the feed industry. That's because true protein is not a simple, uniform compound. Proteins come in many forms: enzymes, muscle fibers, hormones, plant storage compounds, and all sorts of other complex molecules that make living things tick. Each protein molecule is composed of amino acids linked together like beads on a string. Each individual protein contains its own unique set of amino acids which makes it different from all other proteins. This makes true analysis incredibly difficult, because what do you analyze for? All proteins, however, *do* have one thing common: nitrogen. All amino acids contain nitrogen, so it follows that all proteins contain some nitrogen.

So . . . chemists sat around one night in the late 1800s and figured out a simple workaround solution: instead of testing for protein (which would drive you crazy), test for nitrogen. It's the chemical equivalent of using baling twine to hinge a gate; it may not be perfect, but it works. A nitrogen assay is easy (for chemists). It's done with glass beakers and bubbling liquids, things that even high school chemistry labs can do. (Well . . . they did it that way once upon a time). In most commercial labs, this assay is now automated and high-tech, but the principles are still the same. All you have to do is figure out how to convert this nitrogen value into protein. Simple.

This is how it's done: A laboratory runs a fairly simple nitrogen test called a *Kjeldahl Analysis* (pronounced "kell-dall"), so named after the Danish chemist who developed it around the time of dinosaurs. This assay gives a very accurate and precise value for nitrogen. And here's the key conversion: most proteins, on average, contain 16% nitrogen. Therefore, to get to 100%, you simply multiply your nitrogen value by 6.25 (translation: 16 x 6.25 = 100).

Let's take a familiar example: corn. Say we send some corn samples to the lab. They test it and obtain a nitrogen value of 1.62%. They then multiply that value by 6.25 and come up with a calculated protein value of 10.1%. This is the number they report back to you as "crude protein." You can't call it real protein because the lab didn't test for real protein; it only tested for nitrogen.

Same for hay. You carefully take a set of core samples of your hay, send it in, and get a "crude protein" value of 12.3%. This means that your hay actually tested at 1.97% nitrogen (1.97 x 6.25 = 12.3).

Why all this fuss? Well, if you don't know what they mean, crude protein values can be somewhat misleading. Remember that *any* nitrogen showing up in the analysis will be arithmetically converted to crude protein. Like urea, which is a useable form of *nonprotein nitrogen*. Even nitrogen that your livestock cannot use will appear as crude protein in the lab report. For example, some nitrogen is bound up in the cell walls of forages — this nitrogen goes in one end of your livestock and comes out the other end, totally undigested. Crude protein down the drain, so to speak.

So watch out this winter. Did anyone make hay in May or June this year? How wet was it baled? In western Oregon last spring, things were a little wetter than normal. Not as torrential as in some places in the Midwest, but rainy enough that salmon spawned in the lower forty. Some of our hay was baled a little wet. What about yours? Okay, maybe your hay wasn't wet enough to burn your barn down, but next winter when you feed that hay, check its color. Does it look a little (or a lot) brown? That brown color may be a symptom that some of your hay has *caramelized* — the same process that you use in the kitchen to make those delicious candies called *caramels*. This browning often occurs in damp hay, and it means that some of the hay's protein has chemically combined with some of the hay's carbohydrates to form a type of caramel, which is totally indigestible to your livestock. Yet, because this caramel contains nitrogen, the lab will detect this nitrogen and dutifully report it as "crude protein."

That's why you need to remember that the word "crude" in crude protein is actually not crude at all. It's a flag to remind you to evaluate that number very carefully.

First Published: November 1993

Author's Note: This was the first article I wrote for my *"From The Feed Trough..."* monthly column. I took one of the most obvious topics — crude protein — and ruminated about it. This may be a pretty basic topic, and later articles go into some of these details in much greater depth, but it still represents a fundamental principle in nutrition. Well, what do you expect from a ruminant nutritionist?

Dissecting Protein

We all use the term *Crude Protein* to describe nutritional value without much thought about it, as if Crude Protein were something that has been around forever, like "pi" or "I Love Lucy." But if you look carefully at the most recent NRC reference tables for beef and dairy cattle — you'll see that these tables go well beyond Crude Protein. They dissect protein into its component parts to describe feed quality and animal requirements. What's going on here?

Here's some background: Crude Protein (CP) is called "crude" because its official assay analyzes only for nitrogen, not for real protein. This is a practical approach because it's much easier to analyze for nitrogen rather than for true protein, and all proteins are composed of amino acids which contain nitrogen. Since, on average, most proteins contain 16% nitrogen, we simply analyze for nitrogen, multiply that value by 6.25 (because 6.25 x 16% = 100%), and call the resulting number *Crude Protein*. For example, a feedstuff containing 2.1% nitrogen would be listed on a feed report as 13.1% Crude Protein.

But should the world of protein be so easily defined? In reality, protein nutrition for sheep and cattle is far more complex than a factor of 6.25. Lots of things happen to protein and nitrogen when animals consume feed: microbes in the rumen and large intestine convert some of that protein into their own nitrogen compounds; nitrogen is recycled between generations of microbes; some feed nitrogen ends up in nonprotein forms that are digested quite differently than real protein; and saliva, which contains nitrogen, effectively recycles excess metabolic nitrogen back into the rumen. The simple number for Crude Protein accounts for none of these conversions, and more importantly, it does not accurately predict the amount of true protein that a ration supplies to the animal for absorption.

Nutritionists have known about these issues for more than 70 years. The scientist H. H. Mitchell actually discussed some of them in a landmark paper published in 1929. Back then, however, we did not have the appropriate laboratory assays or enough understanding of rumen ecology to solve this problem.

But over the past 25 years this has changed, and nutritionists have developed the concept of *Metabolizable Protein* (MP) to evaluate protein nutrition. The concept is fairly straightforward: MP is the protein that an animal actually absorbs across the gut wall, while CP is the nitrogen that first enters the GI tract when an animal consumes feed. The trick is knowing how to convert CP to MP. This new protein system does just that.

First, Crude Protein is partitioned into two fundamentally different classes of protein:

Class #1: Protein that passes intact through the rumen without destruction or alteration by rumen microbes. Since this type of protein is not degraded in the rumen, it's called *Undegradable Intake Protein* (UIP). (Get used to strange acronyms in protein nutrition. Sometimes I think that they were conceived by the same folks who defined an elevator as a "vertical access device.") UIP is generally insoluble under rumen conditions and therefore cannot be fermented by the rumen microbes. The popular press often calls this type of protein *bypass protein.*

Class #2: Protein that enters the rumen but then is degraded by rumen bacteria *before* it can pass into the lower tract. This type of protein is called *Degradable Intake Protein* (DIP). The rumen bacteria first dissemble these proteins into amino acids and then use the nitrogen for their own compounds. Most of these bacterial compounds are true proteins, but some of these compounds are nucleic acids and cell wall molecules that are not digestible in the lower tract. Therefore, there is a certain amount of digestive inefficiency with all feed proteins classified as DIP, because some of their nitrogen is converted to indigestible compounds that are unavailable to the animal.

Second, each class of protein (UIP and DIP) has different digestibility characteristics in the lower tract. Remember, the digestibility of a protein is the amount that is actually absorbed across the intestinal wall into the blood. In general, the digestibility of UIP is approximately 80%. That 80% portion is called *Metabolizable Feed Protein.* The remaining 20% of the UIP is composed of indigestible nitrogen compounds, such as nitrogen bound up in lignified plant cell walls or caramel polymers of heat damaged protein (= Maillard products, which occur in wet hay after it heats).

In contrast to UIP, the digestibility of DIP is more complicated because of the actions of rumen microbes. Our calculation here is a two-step process: (1) First, we estimate the total amount of protein made by the bacteria, which is called *Microbial Crude Protein.* This amount depends on microbial yield, which depends on the amount of fermentable material in the feed. Higher amounts of fermentable material usually support higher levels of microbial yield. The amount of fermentable material in a feed is conveniently estimated by the feed's TDN value. In general, the yield of microbial crude protein is approximately 13% of the TDN, although this percentage can be smaller for diets containing lower levels of fiber. (2) Second, we need to adjust the microbial crude protein for the losses associated with bacterial action. We know that rumen bacteria use some of the DIP to create compounds that are not digestible in the lower tract, such as microbial nucleic acids and nitrogen compounds in bacterial cell walls (compounds like *diaminopimelic acid* — DAPA). These indigestible compounds average approximately 36% of the total microbial protein, and we account for them by multiplying the value of microbial crude protein by 0.64. The resulting number is called the *Metabolizable Bacterial Protein.*

And finally, we add the values of Metabolizable Feed Protein (from UIP) and Metabolizable Bacterial Protein (from DIP) to gives us a number called the *Total Metabolizable Protein in the Feed* — the total amount of real protein absorbed by animals from the lower tract. This number is the MP value that you see in reference tables.

That's a lot of calculations. But look around — those MP values may start showing up in more places than just reference tables.

First Published: August 1999

Author's Note: Although this topic may seem confusing and maddingly academic, it's really not. And I'm not trying to play the role of a fortune-teller. Nutritional science — particularly ruminant nutrition — has been moving in this direction for years. In fact, the newly-published NRC reference book *Nutrient Requirements of Small Ruminants* (2007) discusses protein *only* in terms of UIP, DIP, and MP, specifically in grams per day of each category, with the term Crude Protein used only as an adjunct concept. We might as well get used to the new acronyms — they will be with us for quite awhile.

Where Does the Protein Go?

Have you ever wondered why a forage containing 26% protein in early April contains only 8% protein when it ends up in your barn in June? Of course you learned in school that forages always lose protein as they mature — but that's like receiving a note from the Department of Redundancy Department. We *know* that mature forages are low in protein — 200 tons of 8% hay in your barn proves it — so being told that forages lose protein as they mature isn't exactly new information. The real question is: What *precisely* happens to that protein as the plant matures? Where does this protein *go*?

First, let's clarify the term *protein*. The protein value that everyone throws around *does not mean true protein*. The proper term for this value is *Crude Protein;* crude being an appropriate designation because the official laboratory procedure is actually a test for *nitrogen*, not protein.

There is a good reason for this: Proteins are large, complex molecules, and plant tissues contain thousands of different types of them. If you wanted to know the level of plant protein for nutritional purposes, which protein would you analyze? There are indeed some laboratory assays available for true protein, but these assays are complex, expensive, and slow. To solve this problem, nutritionists decided to test for something that *all* proteins contain: *nitrogen*. Remember that all proteins are composed of amino acids, and each amino acid contains nitrogen. And because *on the average*, proteins contain 16% nitrogen, multiplying the nitrogen value by 6.25 gives a reasonable value for Crude Protein (16% x 6.25 = 100%). For example, a 1.0% nitrogen assay equals a crude protein value of 6.25%; a 2.0% nitrogen assay equals 12.5%; and so on.

Assaying for nitrogen, however, implies a built-in discrepancy: We use the 6.25 conversion factor based on the assumption that all the nitrogen is contained in true protein. Unfortunately, this assumption is not 100% valid. In addition to true proteins, plant cells also contain nitrogen in many *nonprotein* molecules — like free amino acids, unconverted nitrates, nucleic acids (DNA and RNA), nitrogen associated with plant fiber, and nitrogen-containing caramel-polymers in heat-damaged hay or silage. Collectively, these nonprotein nitrogen compounds are called, logically, *Nonprotein Nitrogen* (NPN). The good news is that most forage NPN is soluble and generally usable by ruminants, so in the end, the umbrella term "Crude Protein" still makes sense in practical situations.

But in this article we are trying to dig deeper, and since plant nitrogen occurs in so many forms, focusing our discussion on the term "Crude Protein" can mask subtle changes in nitrogen levels that occur in plant tissues as they mature. We really should focus on the amount of *nitrogen* rather than the amount of *protein*. So here we go . . .

Let's look at that grass plant in early spring, when it clocks in at 26% crude protein (= 4.16% nitrogen). This plant's vegetation is composed entirely of deep green leaves that seem to rise directly from the ground, with almost no stem. The leaf cells are extremely active metabolically. These leaf cells contain enzymes and other proteins that do the heavy metabolic work: photosynthesis; converting absorbed nitrates into protein; creating new plant tissue; etc. Young grass blades also have NDF values of less than 55%, which means they contain relatively low amounts of cell wall. (NDF = *Neutral Detergent Fiber*, a good estimation of total cell wall.)

Because these plants are growing on well-fertilized soil, their roots absorb nitrates at prodigious rates and transport them upwards to the leaves. Leaf cells, in turn, convert these nitrates into amino acids and then into proteins. Some amino acids, however, can't be converted into true proteins fast enough, so they tend to accumulate in the leaves, and some nitrate molecules can't be converted into amino acids fast enough, so they also accumulate. The resulting accumulation in the leaves is a buildup of NPN, particularly as nitrates and unprocessed amino acids.

How much? Well, young grass leaves may contain one third of their nitrogen as NPN. In concrete numbers, 100 lb of dry matter of our young 26% grass contains 4.16 lb of nitrogen (= 26 ÷ 6.25), with 2.79 lb as true protein and 1.37 lb as NPN.

Now let's look at this same grass when it gets stacked in our barn as 8% hay (= 1.28% nitrogen).

By the time we cut it for hay, these plants have already headed out. We harvest a good yield of leaves, but also lots of seedheads and stems. Physiologically, the stems of the vegetative tillers have elongated like miniature bamboo shoots, and reproductive stems have also developed. All stems are high in fiber and very low in nitrogen, since their primary function is structural. Stems act as scaffolding, stretching out the plant and holding up the seedheads. Stems don't contain much active photosynthesis and therefore have little need for much protein. The leaves of these grass plants also change. As they mature, their fiber content increases. Their cell walls thicken and also become impregnated with *lignin*, a non-nitrogenous compound that increases fiber structural strength but effectively reduces the proportion of nitrogen in the cells. The NDF value of these plants rises above 65%, and the level of true protein in the leaves is diluted by that extra cell wall.

In addition, the maturing plant redirects its metabolic energy into seed formation and slows down leaf growth and other metabolic activities. Less nitrate is carried to the leaves, and the backlog of leaf NPN is used without replacement. In these mature leaves, NPN may represent only 15% of the total nitrogen rather than the 33% level earlier in the growing season.

Back to the numbers: 100 lb of dry matter of our 8% hay contains only 1.28 lb of nitrogen, with 1.09 lb as true protein and 0.19 lb as NPN.

Let's review the entire process.

If we track 100 lb of forage dry matter from early spring to hay harvest, the total amount of nitrogen falls by 2.88 lb (= 4.16 – 1.28). The nitrogen contained in true protein declines from 2.79 lb to 1.28 lb — a loss of 1.51 lb (54%) that can be attributed primarily to dilution of plant material by the increased amounts of fiber. While the number of protein molecules don't increase in the maturing leaves, the amount of fiber increases dramatically, so as additional leaf tissue is created, the existing amount of protein is diluted further and further by fiber that doesn't contain much nitrogen.

Also, in the leaves, the nitrogen contained in NPN drops even further, from 1.37 lb to 0.19 lb, a decline of 1.18 lb, or 86%. Maturing leaves don't need lots of excess amino acids or nitrates hanging around the cells. Nature is ruthlessly economical, and excess inventory is metabolically expensive.

Then of course, add the stems, which increase yield of the mature plants but contain little nitrogen.

So, where does all the protein go? Actually, we should rephrase the question to *where does all the nitrogen go*? In short, to maturity, to fiber, to structure, and to growth. Every year we race against the relentless process of maturity. And if we want to capture higher levels of nitrogen, we really do need to cut earlier.

First Published: June 2004

Author's Note: The idea for this article came from a straightforward question by a member of a local rancher study group. At one meeting, during a discussion about harvesting hay, he asked where does all the protein go? So here is the answer.

A Better Way than Crude Fiber

Oh, the wonderful world of fiber!

It's a world full of confusing, shifting, mislabeled, and misunderstood compounds and structures. It's a world so important that it keeps doctors, nutritionists, federal regulators, timber salesmen, and graduate students busy and off the streets. It's also a world that feeds our livestock. Because fiber is the stuff that plants are made of.

Fiber is another name for cell wall. A bit of chemistry here: plant cells are not quite like your cells and my cells (in case you haven't looked lately). Each plant cell is surrounded by a relatively stiff wall-like structure called, not surprisingly, a *cell wall*. These cell walls give plant cells strength and stability — kind of like an external skeleton, or the scaffolding surrounding a building under construction. This structural integrity is why Douglas Firs can grow to 300 feet and cattle, thankfully, cannot.

What is the most abundant carbohydrate in the world? Not table sugar, or starch, or even white bread. The most plentiful carbohydrate in the world is *cellulose*, a type of fiber that is found in most plant cell walls. Cellulose is usually combined with other types of cell wall compounds to form complex fibrous structures. But there is one special instance in nature in which cellulose is found in a pure form: We call it *cotton*.

But cellulose is only one type of plant fiber. Other types of fibrous substances in our forages are hemicellulose, which is chemically similar to cellulose but is spelled differently; lignin, which is not digestible at all; pectin, which is typically found in fruits; and silica, which is often found in rice straw and sand. Silica is not very nutritious.

One basic rule: the nutritional value of a forage is primarily determined by the amount and type of fiber in that forage. The higher the fiber . . . the lower the digestibility. Also, the higher the fiber . . . the lower the intake. Some plants naturally have more fiber than others. For example, grasses usually contain more fiber than legumes. Nonetheless, the basic relationship between fiber and digestibility always remains the same. Also, as forages grow and mature, their fiber levels increase and the digestibility of this fiber decreases. Translation: older, more fibrous plants are poorer quality feeds (Not exactly a surprising concept). Since fiber levels are so closely related to nutritional value, wouldn't the accurate knowledge of fiber values help us make better decisions about forages, like when we try to buy hay? But there has been a long-standing problem with this . . .

Historically, fiber has always been labeled *Crude Fiber*. Actually, this label is quite misleading. For many years, every nutritionist knew that the crude fiber assay was inaccurate and, well, crude. A crude fiber number told us very little about the real nutritional value of a forage. The crude fiber assay missed a lot of fibrous things in plants and incorrectly identified other things. Although crude fiber was pretty good at estimating the cellulose content of plants, it completely missed lignin, which is the most critical factor in changing a plant's digestibility It also missed nearly all of the pectin and some of the hemicellulose. Nonetheless, the crude fiber assay was the only game in town, and it became the legal standard for labeling animal feeds. In fact, you'll still see it on most animal feed tags. And the next time you visit a supermarket, look at a package label — crude fiber values are also listed on the labels for human foods.

Which brings us to one wild and crazy guy. Back in the 1950's, a young scientist from a dairy farm near Seattle went to work at the USDA nutritional lab in Beltsville, Maryland. His name was Peter Van Soest. Peter danced to the beat of a different drummer. He stayed in his lab for days without a break, slept at odd hours, and generally did not obey too many office rules. In short, he was not your average government employee.

Peter was, however, a supremely gifted scientist, and he clearly recognized that crude fiber was a misleading value. He devoted his research to designing a better fiber analysis — and he succeeded. From his brilliant work came two assays that have, over the years, revolutionized our understanding of livestock nutrition. Those assays are called *Neutral Detergent Fiber (NDF)* and *Acid Detergent Fiber (ADF).*

Those assays have completely changed how we predict forage digestibility, intake, and growth. Using Peter's methods, we can now derive forage digestibility values that are biologically coherent, and we can better predict how forages change nutritionally as they grow and mature. Think about Peter the next time you see a forage report, or make hay, or watch your animals graze pasture.

And that, as they say, is the rest of the story.

First Published: March 1996

Author's Note: Peter Van Soest is a shining example of one person making a difference. I can't overstate the importance of his work. Peter's fiber assays have become the mainstays of ruminant nutrition. They are the building blocks of sophisticated mathematical models for predicting nutrient requirements, as well as tools for understanding the true dynamics of ruminant and non-ruminant nutrition. I took courses from him when I was in graduate school. There was no one else like him. Thanks, Peter.

Fiber Site Map

In 1970, when Keith Goering and Peter Van Soest published their USDA Handbook No. 379 "*Forage Fiber Analyses (Apparatus, Reagents, Procedures, and Some Applications)*," they turned the nutritional world on its ear. In one giant sweep, their new analytical procedures changed the way we look at fiber, forages, and nutritional value, as well as rumen function, predictive equations, and the entire array of nutritional aspects that guide our dietary decisions.

When laid out in a neat diagram, these procedures look like a logic tree with trunks and branches. Or, in current lingo, they look like a website diagram: main pages with links to subpages. Many well-designed websites include a "site map" where visitors can view the entire structure all at once and see where information can be found. So, let's do the same thing with forage analysis — we'll take a brief tour of the fiber site map.

First, some terminology: The key word is *Fiber*. Although fiber is often a buzzword in the popular press, and lots of folks may think they are familiar with it, the reality is quite complicated. Fiber is not fiber is not fiber. Plants contain different types of fiber, and each type functions and reacts differently. The main fiber types are cellulose, hemicellulose, lignin, pectin, and cutin, as well as a few other compounds that nutritionally act like indigestible fiber, such as silica and the heat-damaged protein-carbohydrate polymers called *Maillard Products*.

Briefly, *Cellulose* is a huge fibrous thread containing individual glucose molecules strung together in long chains — very long chains. *Hemicellulose* is a branched fibrous compound containing straight chains and also cross-linkages made of glucose molecules and other sugars. *Lignin* is a radically different type of fiber — a stiff, rigid compound containing benzene rings, which are ultra-stable 6-carbon circular structures with shared double bonds. Benzene rings are so stable that, on one hand, they act as admirable "stiffening" components in fiber, but on the other hand, an animal's digestive tract contains no intestinal enzymes strong enough to break these rings apart. This makes lignin virtually indigestible.

Two other common types of fiber are *Pectin*, a complex fibrous compound found in high concentrations in some fruits and seed hulls, and *Cutin*, a waxy indigestible substance that nutritionally acts like a fiber. Pectin is actually quite digestible under some conditions, while cutin is essentially indigestible.

One thing to remember: I'm keeping these characterizations relatively simple. In reality, fiber molecules are incredibly complex and variable, and these basic types of fibers are really *families* of compounds. But let's reserve that complexity for the large textbooks. In this article I'll give you a general overview of fiber — something to hang your hat on.

Now for the analytical procedures. The ancient, popularized method of determining fiber is called *Crude Fiber (CF)*, and it is an analysis of disappearance. A forage sample is first boiled in a weak acid solution (dilute sulfuric acid) and then boiled again in a weak alkali solution (dilute sodium hydroxide). The residue, which contains fiber and minerals, is weighed and then burned in a furnace, destroying the fiber and leaving the minerals. The material that disappears during the burning is called *Crude Fiber.*

The main problem with Crude Fiber is that nutritionally it is a mess. The CF number does *not* represent the total amount of fiber in a feed or forage. The CF procedure systematically misses some of the hemicellulose and most of the lignin. Since hemicellulose is a major component of most fiber, and the amount of lignin directly affects the digestibility of the total fiber, the CF value does *not* accurately portray the nutritional characteristics of a forage. In other words, Crude Fiber is too crude to be useful.

In contrast, most of the fiber procedures of Goering and Van Soest are analyses of *detection,* not *disappearance.* At each step, the material *remaining* in the container is weighed. Part of the brilliance in designing this system was that these researchers relied on special detergents to dissolve the components they didn't want to measure. They chose two different detergent solutions — one with a neutral pH (called *Neutral Detergent*) and one with an acidic pH (called *Acid Detergent*). In a sense, this detergent system of fiber analysis is kind of like the instructions for creating a stone sculpture of a horse — you start with a large rock and then chip away everything that doesn't look like a horse.

Let's begin with the most basic separation method — boiling the sample in a Neutral Detergent solution. The resulting residue is a fibrous mass logically called *Neutral Detergent Fiber (NDF).* This is the NDF number that you see on forage test reports.

This NDF value represents most of the true fiber in a sample. It includes all the cellulose, hemicellulose, lignin, cutin, as well as some other nutritionally indigestible compounds like Maillard Products and silica. (Maillard products are the gooey black caramel polymers that occur in heat-damaged hay and silage. Silica is, well, the mineral in sand. Many plants, like rice straw, naturally accumulate silica in their cell walls to improve fiber strength. But silica has zero nutritional value, so it's good to quantify it in the fiber portion of the analysis). NDF, in fact, is the most accurate measurement of the amount of *cell wall* in a sample. Cell wall, as you may guess, contains nearly all the fiber in a plant. True, NDF does not include pectin, but pectin is a highly digestible fiber, while NDF represents the types of fiber that can have great variation in digestibility or are not digestible at all.

NDF is a good number to know because of all the nutritional values in a test report, NDF is the number most highly correlated with *forage intake*. Correlated in a negative way, of course — higher NDF numbers are associated with *lower* forage intakes. This makes sense when you consider that high fiber levels cause a sense of rumen "fill," and that an animal with a rumen full of fiber would be less willing to consume more forage than an animal with a rumen containing less fiber.

If NDF is the *residue* of the first separation procedure, what about the flip side of this analysis? *What about the stuff that dissolves into the neutral detergent solution*? This fraction is also very important nutritionally and represents another strength of the detergent fiber analysis system, so let's discuss it now.

If NDF represents the cell *wall* of plant cells, then the soluble portion in neutral detergent represents the cell *contents*. This soluble fraction, commonly known as *Neutral Detergent Solubles (ND Solubles)*, contains the sugars, starches, fats, pectin, soluble proteins, and nonprotein nitrogen in the feedstuff, as well as some minor items like vitamins and secondary plant compounds. In practice, everything in this soluble fraction is nearly 100% digestible, and the calculation is easy — just subtract the NDF number from 100. For example, if a forage contains 61.2% NDF, then its cell contents is 38.8% (= 100.0 - 61.2).

Perhaps you've heard of something called *Non-Structural Carbohydrates (NSC)* or *Non-Fiber Carbohydrates (NFC)*? These terms are part of the ND Solubles and represent the amount of sugar plus starch in the forage. Sometimes NSC and NFC are used interchangeably, but there are really slight differences that are debated among nutritionists. In essence, NSC is the result of the *direct analysis* of starches and sugars, while NFC is a *calculated value* based on a formula. Let's not go into that controversy here. In either case, NSC is important because it represents the carbohydrates that may ferment relatively quickly in the rumen, and this has implications on acidosis, the metabolism of nonprotein nitrogen in forages, and the new high-sugar grasses. But more on this later.

Oops. Out of space. Well, next time, we'll take apart NDF and its cousin, ADF, and show what these analyses can reveal about forage nutritional value.

First Published: February 2005

Author's Note: NDF and ADF are the touchstones of our modern understanding of ruminant nutrition but in the formal scheme of things, these terms do not appear on feed tags or other official feed descriptions. Crude Fiber, on the other hand, tells us little about the nutritional value of a feed or forage, but it is still the legal fiber term required on feed tags, even after more than 30 years of the NDF/ADF system. It makes one wonder.

More Fiber Numbers

Last month I outlined the general system of the modern fiber analysis and described two main numbers from that analysis: NDF (neutral detergent fiber) and NFC (non-fiber carbohydrates). This month, let's look at the other main fiber number — ADF — and also an array of useful values we can derive from it.

The basic procedure for analyzing ADF in a forage sample is to dry the sample, grind it, and then boil it in a beaker containing special detergent solution that has been carefully adjusted to an acid pH. This solution is logically called *Acid Detergent*, and the fibrous substance remaining after the boiling process is called *Acid Detergent Fiber (ADF)*. This is the ADF value you see on forage test reports.

When you study a forage test report, one thing should be immediately obvious: ADF is always smaller than NDF. The reason is that ADF contains fewer components than NDF. If you remember from last month, NDF represents the entire cell wall of the plant and consists primarily of cellulose, hemicellulose, lignin, cutin, and some minor fibrous compounds. In contrast, the acid in the ADF procedure dissolves the hemicellulose, leaving nearly all the other fibrous components. Which means that, with minor exceptions, ADF contains everything in NDF except the hemicellulose. Which also means that we can easily calculate the amount of hemicellulose in a forage by subtracting ADF from NDF. For example, a forage with 62.5% NDF and 32.8% ADF contains 29.7% hemicellulose (= 62.5 - 32.8). Well okay, perhaps this number won't become a hot topic at your family supper table, but it can be handy when you need it.

One of the beauties of the detergent system is that we can use it to run *sequential* procedures. Which means that we can take a fiber residue like ADF from one part of the system and *then* run procedure X or Y or Z on it, and each of these procedures will give us additional information about the nutritional makeup of the forage.

So let's start with the ADF residue — the fibrous mass at the bottom of the beaker — and test *that residue* for *nitrogen*. This means running the classic

Kjeldahl procedure (pronounced "Kell-dall") involving lots of glassware and boiling liquids. The resulting number is the amount of nitrogen in the sample.

Which means exactly what? If this nitrogen number were derived from the original forage sample, we would multiply it by 6.25 to calculate the amount of *Crude Protein* in the entire forage. But remember, we started with the ADF, not the entire forage. Since we don't expect much true protein in ADF, any nitrogen present is probably bound into compounds with little nutritional value. Therefore, if we want to know the true amount of crude protein nutritionally available to an animal, we must subtract the ADF-Nitrogen from the total nitrogen of the feed. Read on.

Nitrogen in ADF generally occurs in one of two forms, and both are indigestible: *Maillard Products* and *leather*. Leather? Well, not very much. Even if some forages contain lots of tannins, we really don't expect plants to produce much leather (unless, of course, they are leather plants . . . get it?). On the other hand, Maillard Products are rather common. Maillard Products are the gooey, indigestible polymers formed from carbohydrates and proteins during heat damage — which occurs in wet hay or poorly-made silage. The nitrogen in ADF is called, logically, *Acid Detergent Insoluble Nitrogen* (ADIN). Multiplying *this nitrogen value* by 6.25 gives a number that appears in forage reports as *Heat-Damaged Protein*. If this number is large compared to the total Crude Protein, it means that the forage experienced a significant amount of heat damage, and the total protein value should be reduced by the amount of heat-damaged protein. For example, if your hay sample contains a total crude protein value of 13%, and the forage test shows 2.5% heat-damaged protein, you should use 10.5% in your ration-balancing calculations, not 13%.

Another procedure with ADF is to burn the ADF residue in an oven and then bathe the resulting ash in a strong solution of *hydrogen bromide* (note: do *not* try this at home!). Hydrogen bromide is an *extremely* strong acid that dissolves all minerals except silica. Silica is abundant in the soil, and some plants incorporate it into their fiber to improve rigidity. Although silica is not digestible, it can still influence nutrient digestibility, so the amount of silica in a forage is a good thing to know. But silica can also come from soil contamination, so the silica number should be viewed with a grain of salt.

Now for a matched pair of procedures that assay for three important types of fiber: cutin, cellulose, and lignin.

Cutin is a complex, waxy, indigestible substance found in many leaves and stems. *Cellulose* is a very large molecule that is a polymer chain of glucose units, strung together in long fibrous strands. *Lignin* is composed of rigid ring-structures that are exceptionally stable and strong. So strong, in fact, that lignin cannot be digested by mammals because no digestive enzymes in the rumen or lower tract can break it apart. But in a laboratory, we have access to some chemical reagents that are not exactly found in gastrointestinal tracts.

One such laboratory reagent is *Sulfuric Acid* (H_2SO_4). A concentrated solution of 72% sulfuric acid will dissolve cellulose. If you don't believe me, consider this: what happens when you are working on a car and spill battery acid on your jeans? Your jeans become holey, right? Why? Because strong acids dissolve cellulose, and jeans are made of cotton, and cotton is . . . uh . . . cellulose. (And that's why you should wear wool pants when you work with car batteries).

The other laboratory reagent is *Potassium Permanganate* ($KMnO_4$). This is an extremely strong *oxidizing agent* which can break up those ring-structures of lignin and dissolve it.

Let's use these compounds in two different sequential procedures. In both procedures, we begin with the ADF residue and *then* run an NDF analysis on it. The NDF step removes some interfering factors and leaves a residue that is analytically cleaner for our purposes than either ADF or NDF alone. Trust me on this.

Now for Sequential Procedure #1: We first treat this two-step residue with 72% sulfuric acid, which removes the cellulose and leaves a substance containing both lignin and cutin, which we weigh. Then we treat *this* residue with potassium permanganate which dissolves away the lignin. Now all we have left is the cutin. But think for a moment — we have just derived the values for *two* types of fiber. We know the amount of cutin because we can weigh it in the beaker, but we also know the amount of lignin. How? By subtraction. The pre-permanganate residue contained *both* lignin and cutin, but the final residue contained *only* cutin. Therefore, the substance dissolved by the permanganate was the lignin.

Still following me? Here's Sequential Procedure #2: We do the same thing as outlined in the previous paragraph, *but in reverse order.* First the potassium permanganate, then the sulfuric acid. As before, the final compound in the beaker is cutin, but this time the subtraction gives us the amount of *cellulose.* Okay — you can diagram the details by yourself.

Fiber analysis for livestock nutrition has certainly come a long way from the crude old days of crude fiber. But ironically some things don't change: The fiber value listed on human foods is still *crude fiber*. Just walk into any supermarket and read the package labels. No numbers for NDF, ADF, heat-damaged protein, or cellulose. It's the law. It's also something to ruminate on.

First Published: March 2005

Author's Note: In nutrition, the Detergent System of fiber analysis stands alone in its breathtaking importance. With these procedures, nutritionists for the first time accurately identified fiber components with nutritional integrity, which paved the way for understanding basic nutritional concepts like rate of passage, particle size, rumen fermentation, bypass protein, and other issues that we consider routine today.

Cellulose Threads

Here's a question for you: of all the common feedstuffs in the world, which one is the most abundant? Hint: it's not corn or wheat. Think of forages. Think of fiber. Think of *cellulose*. All the tons of plants that cover the land; all that forage with stems and stalks; all that plant fiber — it all contains cellulose.

In the broadest sense, cellulose is sunlight captured in a fibrous form — a compound that is universal, nutritious, and completely renewable. But cellulose is more important than just something to eat. It's also the source of derivative compounds that are used throughout the world in all sorts of practical and surprising roles. Let's talk about cellulose and its derivatives.

Exactly what is cellulose? This may get a bit complicated, but let's begin with its basic building block: the simple sugar molecule called *glucose*. Glucose contains six carbon atoms with five of these atoms arranged in a ring and the #6C branching off from the #5C. Five of these carbons are also each linked to a hydroxyl group (-OH). Glucose is a critical molecule for animal and plant metabolism — it's the primary source of energy for most metabolic processes as well as the starting molecule for synthesizing thousands of other biological compounds. Our interest here, however, is in how it combines with other glucose molecules.

If we connect *two* glucose molecules by linking the #1C of one glucose with the #4C of the other glucose, we will create a two-glucose molecule called *maltose*, which is a disaccharide (*di* = two, *saccharide* = sugar). Two other disaccharides are quite familiar to us: the milk sugar *lactose*, which contains one glucose and one galactose, and the supermarket sugar *sucrose*, which contains one glucose and one fructose. We all know sucrose by its sweet common name — *table sugar.*

If we connect *thousands* of glucose molecules in a straight-line chain, like a string of pearls, with each pair of glucose molecules connected with a C1–C4 linkage, we have a glucose polymer called a polysaccharide (*poly* = many). But there's a rub: the C1–C4 bond can exist in *two* possible arrangements: (1) when the C1 bond points downwards, the bond is called an *alpha* linkage (α-linkage), and (2) when the C1 bond points upwards, the bond is called a *beta* linkage (β-linkage). I know that this is beginning to sound rather academic, but just hang on. Things now get interesting.

If thousands of glucose units are linked in a chain by *alpha* 1–4 bonds, the molecule is called *starch*. Plants use starch as their primary storage compound, and all of our common food seeds — corn, wheat, barley, oats, etc. — are really just packages of starch. Animals synthesize a smaller version of this glucose polymer as a quick-access source of energy. In animals, this smaller polymer is called *glycogen*.

On the other hand, if thousands of glucose units are linked in a chain by *beta* 1–4 bonds, the molecule is called *cellulose*. Plants use this polymer as part of their structural material for building stems and leaves. The geometry of these beta linkages allows the long cellulose chains to align themselves side-by-side and thus create strong fibrous strands.

Let's return to the nutritional value of cellulose in feeds. Do all those quadzillion tons of cellulose in the world represent a good source of nutrition? Well, yes *and* no. Sure, cellulose contains energy that is *potentially* available to animals, but cellulose is also a very large molecule. Since large molecules can't be absorbed directly across the gut wall, the digestive tract must first enzymatically break them into smaller units that can be absorbed and transported into the blood. But there is a slight problem: *no mammal (or bird, reptile, insect, etc.) has the digestive enzymes that can break those beta 1–4 linkages.* Which means that mammals are unable to digest cellulose directly.

Whoa! If animals can't digest cellulose, how can it be nutritionally important?

Nature has provided an elegant solution: although mammals don't have the enzymes to digest cellulose, certain bacteria *do* have these enzymes. In fact, cellulolytic bacteria thrive quite nicely by using cellulose as their primary food source. By capitalizing on this feature, evolution has come to the rescue. Over millions of years, some animals evolved a specialized pouch in their digestive tracts to house these bacteria. This pouch — a fermentation sac — can either be located at the front end of the digestive tract, where it is called the *rumen*, or at the far end of the digestive tract, where it is called the *large intestine* or the *cecum*. In these fermentation sacs, the cellulolytic bacteria happily do their thing and in turn, produce nutritional products that can be absorbed and used by the host animal for its own metabolism. Thus, the ruminant or horse or alpaca or any animal housing these bacteria (including, to a small extent, humans) can obtain nutritional value from cellulose.

Actually, mammals aren't the only animals who evolved fermentation sacs for cellulolytic bacteria. Termites have done it too, and for exactly the same reason. Although termites may not be good at chewing their cud, most of us can testify to their success at digesting cellulose. And I'm also thinking . . . it's a good thing that cows and sheep evolved to eat grass instead of houses.

Back to cellulose. When we analyze feeds in a nutrition laboratory, we assay plant fiber as *NDF* (Neutral Detergent Fiber). In nature, plant fibers are very complicated structures that contain a number of different fibrous substances. Cellulose is only one component of this fiber matrix, although it is generally the largest one. Other types of common fiber molecules include *hemicellulose*, a complex cellulose-like molecule with lots of side-chains, and *lignin*, a strong indigestible compound that acts to reinforce the entire fiber structure. In nature, cellulose occurs as an integrated part of a complex package, rather than as a stand-alone compound.

With one notable exception: *cotton*. The cotton plant produces a fruit called a *boll*, which contains seeds in a meshwork of cotton fibers. These cotton fibers are essentially pure cellulose. Which is why, I think, that nutritionists sometimes look at a cotton shirt as a collection of glucose chains rather than as a fashion statement.

But here's a thought: if the polymer structure of cellulose is so utilitarian, you'd think that nature would have used it in other places in addition to plant fibers. Well, it has. We've already discussed starch and glycogen, which are glucose polymers built with alpha 1–4 linkages. Another variation occurs in insects.

As I mentioned earlier, every glucose molecule has five carbons linked to hydroxyl groups (–OH). When a glucose molecule is bound in cellulose, two of these carbons are automatically occupied by the glucose-to-glucose linkage (#1C and #4C), and therefore they can't be linked to any hydroxyl groups. But that still leaves three carbons that *do* have hydroxyl groups.

Now, if we take the #2 carbon in every glucose unit and replace its hydroxyl group (–OH) with something called an *acetylamine group* (–NHCOCH3), the resulting polymer is called *chitin*. Like cellulose, chitin molecules can align themselves side-by-side and form tightly condensed flexible sheets. But chitin can be harder and stronger than cellulose because the acetylamine side-chains allow better cross-linkages between the parallel strands. Certain insects and fungi use a flexible form of chitin as their outer layers. But some organisms combine chitin with various proteins to make a molecule that is extremely tough and impenetrable. It's this type of chitin that forms the hard exterior shells of insects like beetles and also crustacea such as lobsters, shrimp, and crabs.

Cellulose, cotton, starch, glycogen, and chitin — natural variations on a glucose theme. Next time, we'll explore some man-made variations on this theme and see how these variations have changed the world.

First Published: January 2007

Author's Note: Before plastics, before space-age materials, there was cellulose. Nature has used it extensively. Cellulose is really sunlight in glucose polymer form. Now *that's* something to ruminate about.

Fiber that Changed the World

Last month I described some chemical details of cellulose — the plant fiber that comprises the most abundant feedstuff in the world. Cellulose, cotton, and chitin are all compounds found in nature. But man doesn't leave nature alone. Over the years, we've fiddled with cellulose to create a whole raft of derivative compounds, some with surprising qualities. In fact, some have changed the world, or at least significant parts of it.

First a short review: cellulose is a polymer of glucose — a long chain of 6-carbon glucose molecules strung together like boxcars in a freight train. In each glucose subunit, two carbon atoms are linked to adjacent glucose subunits with specialized *beta 1–4 linkages,* and three carbon atoms are linked to hydoxyl groups (-OH). So, starting with this natural polymer, here's how we make cellulose derivatives: by substituting other things for those hydroxyl groups. What happens? It depends on the properties of those other things, and whether we replace all or only some of the hydroxyl groups.

Let's start by replacing hydroxyl groups with nitrate groups. For example, by adding nitric acid to cellulose to create something called *nitrocellulose.* Folks first did this commercially in the 1840s, and there is a bit of a legend about it. The story is that in 1845, a German-Swiss chemist named Friedrich Schönbein was experimenting with various acids in his kitchen (apparently, scientists worked at home in those days), when he accidently spilled a bottle of nitric acid. His wife was out, so he quickly grabbed her cotton apron, mopped up the mess, and hung the apron near the stove. After a short time, the apron dried out. Then it blew up. Mr. Schönbein had unwittingly discovered *guncotton,* a nitrocellulose compound with nearly all its hydroxyl groups replaced by nitrates. History didn't record, however, Mrs. Schönbein's reaction to this explosive discovery. I suspect that she was not amused.

Guncotton looks like cotton but packs more explosive power than black powder while producing less smoke. In fact, over the years, variations of guncotton have been marketed as smokeless powder. But guncotton has limited use because it's also *very* delicate and unstable — it has the unnerving characteristic of exploding at the slightest disturbance. As you can imagine, guncotton was not a favorite substance of insurance companies or manufacturing plants, and it is not used much today. By the way, gun*cotton* is not the same as gun*powder*, a substance made from saltpeter (potassium nitrate), charcoal, and sulfur. Gunpowder *is* used today, but that is a different story.

However, if we add nitrates to cellulose under controlled conditions so that we replace only *some* of the hydroxyl groups, we create a different type of nitrocellulose compound called *pyroxylin*, which is the starting point for a very interesting array of useful compounds. For example, if we dissolve pyroxylin in various solvents and add a plasticizer and a pigment, we create beautiful lacquers which are used to finish guitars and other fine wooden items.

If we dissolve pyroxylin in ether and ethanol, we create a compound called *collodion*, which was first discovered in 1848. Although collodion is toxic and flammable, it dries to a waterproof film that has been used in all sorts of products. Early photographers used collodion as an emulsion for their photographic plates. During the Civil War and for years afterward, doctors used collodion as a wound dressing. Even today, makeup artists in theater and film use collodion to simulate age wrinkles and scars — effects you often see in movies — and scientists use collodion for cleaning telescopes and embedding microscope specimens.

Speaking of the movies, did you know that cellulose had a profound effect on photography and Hollywood? By mixing pyroxylin with camphor, dyes, and other compounds, we create a plastic-like substance called *celluloid*. In the late 1800s, many inventors experimented with celluloid to create a flexible transparent plastic. One of these folks thought that this new material looked like a *film* on the surface of spoiled milk. The word *film* has stuck with us ever since. Nineteenth century photographers coated celluloid with their photosensitive emulsions, and the flexible forms of celluloid became the standard material for early film stock. But those old celluloid films were so flammable (especially when a filmstrip jammed in front of a hot projection bulb) that some early theaters actually lined their film projection rooms with asbestos. Happily those days are past.

The hard varieties of celluloid were originally used as substitutes for ivory in everyday items such as knife handles, cufflinks, and toys. While those have been mostly replaced by modern plastics, this type of celluloid still finds some use today for things like ping-pong balls and guitar picks.

Let's leave the nitrocellulose family and switch gears, chemically-speaking. Instead of adding nitric acid to cellulose, let's add sodium hydroxide and carbon disulfide to cellulose to form a compound called *cellulose xanthate*, also known as *viscose*. This thick, viscous substance has its own interesting history, but for brevity, our real interest lies in two things that are made from it.

If we extrude viscose through a narrow slit, we obtain a very thin, very flexible sheet of clear material called *cellophane*. Everyone is familiar with cellophane. A Swiss textile engineer invented cellophane in 1908, and it is now used worldwide for candy wrappings, food packaging, Scotch Tape, and even in the manufacture of fiberglass.

On the other hand, if we extrude viscose through a spinneret — a piece of equipment resembling a shower head with many tiny holes — we'll get thin fibers that can be spun into the fabric called *rayon*. Did you think that *all* synthetic fibers came from petroleum? The next time you put on a rayon shirt or dress, or lie back on a soft rayon bedspread, think about how this fabric began life as a blade of grass instead of being pumped from the earth as oil.

Finally, let's make something *really* different from our cellulose molecule. If we add chloroacetic acid to cellulose in an alkali solution, we'll create a compound called *carboxymethyl cellulose*. Similarly, if we substitute 2-hydroxypropyl ether for chloroacetic acid, we'll make *hydroxypropyl cellulose*. These are both nontoxic compounds with similar properties. The food industry often adds them to recipes to thicken liquids and also to stabilize emulsions such as ice-cream.

And did you know that both compounds are used in artificial tears? Yes, artificial tears. (I'll refrain from mentioning anything about crying over spilt milk). Humans sometimes suffer from a syndrome called *keratoconjunctivitis sicca*, also known as *dry eye syndrome*, and doctors will often prescribe eye-drops containing either of these cellulose compounds. Dogs occasionally suffer from a dry-eye syndrome, and veterinarians recommend similar cellulose-derived products. (Cats, of course, being different, rarely suffer from this problem).

Well . . . we've come a long way from fiber nutrition, but cellulose has indeed affected us all. Explosives, wound dressings, lacquers, photographs, movie films, cellophane, rayon, guitar picks, ping-pong balls, and artificial tears — compounds all owing their origins to photosynthesis — the humble, renewable process where green plant leaves capture sunlight and convert that sunlight into cellulose.

First Published: February 2007

Author's Note: This topic combines fiber chemistry with history — one of my favorite activities. Chemistry tells us what we have today, and history tells us how we got here. But I still wonder about the look on Mrs. Schönbein's face when she walked into that kitchen after her husband blew it up.

Starting Fluid Extracts

Consider the lowly fats. Since they are only a small fraction of most rations, they get no respect. We measure them, note them, neglect them, and move on. Rations include other components that are more compelling, like protein, fiber, and minerals. But fats can contain 225% more energy than carbohydrates. Under the right conditions, they may represent a source of untapped calories for growth and lactation. Such a potential — perhaps we should look at dietary fats more closely.

To understand fat nutrition, we need to discuss two important concepts: (1) how fats are analyzed and (2) what happens to fats in rations, especially their effects in the rumen. We'll cover the first concept this month.

Look at any feedstuff reference table or analysis report, and you won't see a specific value for *fat*. Instead, you'll see something called *ether extract*. Some documents treat this value as synonymous with nutritional fat — but it's not. It's close, but it's not exactly the same. Ether extract is the material in feed that is soluble in ether. True, most fats are soluble in ether, but some are not, and there are quite a few other things soluble in ether that have no nutritional value. Let me explain.

Laboratories analyze fat with a procedure called the *Soxhlet Extraction* (pronounced "socks-let"). This is a technique dating from the 19th century using cleverly-designed glassware, hotplates, and tubes of cold water. The concept is straightforward enough. Essentially, a ground-up feed sample is bathed in liquid ether. Anything soluble in ether dissolves into the fluid. After a series of ether washings and recyclings, the residual sample is weighed, and the difference is called the *ether extract*.

By the way, we should never take this technique lightly, even though it may be old and familiar. Ether is short for *diethyl ether*, which is the chemical name for *engine starting fluid* — the same explosive stuff that we spray into balky engines. The can of starting fluid in my garage has the solemn warning in bold, red letters, "Extremely flammable! Keep away from heat, sparks, or fire!"

Recently I saw an advertisement for a commercial Soxhlet Extraction System that boasts an "explosion-proof hotplate." As they say in many advertisements: do not try this at home.

So, what exactly does the ether extract procedure . . . extract? One thing it extracts is nutritional fat. The basic subunit of nutritional fat is called a *fatty acid*, which is a long-chain molecule composed of carbons and hydrogens with a carboxyl group (–COOH) on one end and a methyl group (–CH_4) on the other end. Fatty acids differ from each other principally in the number of carbon atoms and carbon-to-carbon double bonds in the chain.

From a nutritional perspective, fats that contain 225% more energy than carbohydrates are called *triglycerides*. These are molecules composed of three fatty acids linked together with a 3-carbon glycerol backbone, similar to a 3-horse yolk. Triglycerides are the primary fat molecules in grains and beans and also in tallow, grease, and vegetable oil. For these types of feedstuffs, ether extract does a fairly accurate job of representing their nutritional value.

Let's recognize the one main reason for triglycerides — *energy storage*. Long chains of carbons and hydrogens create molecules that are more densely packed with energy than carbohydrates. This energy-dense design content is no accident. In animals and plants, triglycerides act as physiological bank accounts which store energy until those extra calories are needed. Not surprisingly, we find high levels of triglycerides in certain specialized tissues, such as animal body fat and the portions of some plant seeds (grains).

But forages are important feedstuffs too, and from a fat perspective, forages are quite different from grains. First, most ether extract values for forages are quite low, and second, forages contain few triglycerides in their growing tissues.

Leaves and stems contain no specialized energy storage depots, but still they contain a small amount of ether extract. (Somehow, the image of grass plants with globs of fat hanging from their leaves is rather disturbing). Ether-soluble compounds in forages represent a wide array of functional molecules — galactolipids, waxes, resins, essential oils, pigments, and even chlorophyll. Only some of these compounds have nutritional value, like the *galactolipids*, which are molecules that contain only one fatty acid linked to a single sugar. Galactolipids are digestible, but their energy values are lower than triglycerides because each galactolipid molecule contains a smaller fatty acid percentage than a triglyceride molecule.

Most of the other ether-soluble substances in forages have little or no nutritional value, which means that the ether extract values for forages are not very useful. In practice, this error is usually minimal because the forages generally contain so little ether extract (less than 2% of the dry matter).

One interesting note about fats is that their digestibility values may be wrong. Remember that *digestibility* is a measurement of the amount of material that *disappears* from the gut between the mouth and the manure. To measure this value for fats, we routinely run ether extracts on feed and manure. Using a little arithmetic, we can easily calculate the amount that did *not* appear in the manure. Dividing this amount by the amount in the feed gives us the *apparent digestibility* of the ether extract. For example, if an animal consumes 60 grams of ether extract in the feed, and 6 grams comes out in the manure, we can assume that 54 grams were digested, and therefore conclude that the apparent digestibility of that ether extract is 90% (= 54 ÷ 60 expressed as a percentage).

Unfortunately, as fatty acids move through the gastrointestinal tract, some of them can become linked to calcium or magnesium atoms, making them less digestible. These altered molecules continue traveling downstream through the gut and end up in the manure. Calcium and magnesium salts of fatty acids are called *soaps*. The problem is that soaps are not soluble in ether. Which means that an ether extract analysis of the manure will *not* extract them, which means that the ether extract analysis for manure will result in a *lower* number than it should be, which means that the digestibility for those fats will be *overestimated*. Still following me?

This error may have broader consequences. The traditional formula for calculating the TDN value of feeds involves adding the digestible portions of various nutrient components. The digestible amount of ether extract is multiplied by 2.25 to account for the extra energy in triglycerides. However, depending on the type of feed, triglycerides may not comprise a large percentage of the digestible fats, and if soaps were formed in the gut, then the digestible amount of that ether extract was overestimated. Hmmm.

Starting fluid, soaps, and hotplates that won't explode. There's depth in still waters. Maybe we should read those ether extract values with a little more respect.

First Published: May 2003

Author's Note: To add to the potential confusion, commercial laboratories offer analysis packages that may or may not include ether extract, which they often label as "fat." Perhaps it's a good thing that fat is generally such a small proportion of the dry matter. It's also a good thing to read the submission form carefully to see if fat is indeed part of the package.

Chewing the Fat

I was just looking over my previous articles on fat nutrition and found one published in May 2003 entitled "Starting Fluid Extracts." In one of the first paragraphs, I wrote that I intended to "discuss two important concepts: (1) how fats are analyzed and (2) what happens to fats in rations, especially their effects in the rumen." I said that I would cover the first concept in that month's article — which implied that I would discuss the second concept in a subsequent article, real soon. But now I see that "real soon" has stretched out to years and years. This is a case where I would procrastinate more, if I would ever get around to it . . .

So for all of you who've been waiting anxiously for my follow-up article on fats in feeds . . . here it is. Finally. I'll concentrate on feeding fat to ruminants — sheep, goats, cattle, elk, reindeer, giraffes, and such — and I'll focus on three main concepts: (1) what happens to fats in the rumen, (2) how fats affect rumen fermentation and feed intake, and (3) practical issues about adding fats to feeds.

So, why are fats so attractive in livestock rations? Simple. Ounce for ounce, pound for pound, fats contain far more energy than carbohydrates and protein. Adding fats to rations can significantly improve animal performance or body condition. Or not. There are caveats — big ones — so we need to learn some solid principles before we willy-nilly pour vegetable oil onto our pellets.

Here's the main energy principle from the perspective of animal nutrition: fats contain 9 Calories/gram, while carbohydrates and protein only contain 4 Calories per gram. Multiplying 4 by 2.25 equals 9, which explains why the formula to calculate TDN multiplies the value of digestible fat by 2.25 compared to protein and carbohydrates. The arithmetic, however, can become a bit confusing. Fat contains 125% more calories than the other nutrients, but the coefficient in the formula is 2.25. (Your homework this month . . . explain why this is so.) Whatever terminology we use, these numbers make a very impressive case for using fats to increase the digestible energy level of a diet.

Now some terminology. The term *fat* can refer to a wide array of substances — oil, tallow, lard, fatty acids, all the "glycerides" (tri-, di-, and mono-), etc. All these compounds are nutritionally available, and all contain the same high level of energy, more or less. Analytical labs determine the amount of fat in a feed by the *Ether Extract* procedure. The resulting number represents *all* the compounds that dissolve in ether, but these include nutritional fat *plus* some other non-nutritional substances like waxes, resins, essential oils, pigments, and even chlorophyll. These other substances are not important in grains and most other concentrates because most of the fat of these feedstuffs is actually quite digestible (vegetable oil, etc.). But in forages, which normally don't contain much nutritional fat, these non-nutritional compounds can represent a fairly high percentage of the ether extract value. The good news, however, is that this discrepancy generally has little practical impact because in forages, the ether extract value is usually quite low.

Fat molecules contain *fatty acids*, which are primarily strings of carbons connected by single bonds. (I've discussed these structures in lots of detail in my other articles, so I won't repeat everything here.) If two adjacent carbons are linked with a double bond, the fat is *unsaturated.* If the fatty acid contains two or more double bonds, the fat is *polyunsaturated.* A fatty acid with no double bonds is *saturated.* We also use a shorthand code to describe fats. For example, an 18-carbon fatty acid with two double bonds would be identified as C18:2. A common name for this fat is *linoleic acid.*

Back to the rumen. When a cow or sheep consumes fat, these molecules enter the rumen where they are exposed to rumen microbes in an environment containing excess hydrogen and no free oxygen. The rumen bacteria and protozoa don't metabolize fat directly — they are unable to use the fats for their own energy — but they can alter the structure of some of the fat molecules.

The nutritional fats in grains, vegetable oils, and forages are combinations of saturated and unsaturated fatty acids. Once these fats enter the rumen, bacteria use the excess hydrogen atoms to reduce some double bonds to single bonds. By the time these fat molecules exit the rumen, many have become more saturated.

This phenomenon occurs only in ruminants; it does not occur in animals without rumens. Which explains why the fat of sheep and cattle is more saturated than the fat of pigs. Ruminant fat reflects the saturation of fats in the rumen; pig fat mirrors the relatively unaltered vegetable fats consumed by that animal. Saturated fats tend to have higher melting points than unsaturated fats, which means that at room temperature, ruminant fat tends to be harder than pig fat. Our language has distinct words for these two types: the rendered fat from cattle and sheep is called *tallow*, and the rendered fat from hogs is called *lard.*

One important feature of the rumen saturation process is that it actually occurs in a series of separate steps. Let's focus on a common storage fat in plants called *linoleic acid* (a C18:2 fatty acid with double bonds beginning at carbon #9 and carbon #12). When this fat first enters the rumen, the rumen bacteria reposition one of the double bonds to a different set of carbons. The new C18:2 molecule has double bonds at carbon #9 and carbon #11. Then the bacteria saturate one of its double bonds to form a C18:1 fatty acid with only one double bond at carbon #11. This molecule has the common name *vaccenic acid*. Then the bacteria saturate the remaining double bond to create a fully saturated C18:0 fat called *stearic acid*, which has no double bonds.

The reason I've gone into mind-numbing detail is that the second molecule in this process is rather special. This C18:2 fat is the famous *CLA — conjugated linoleic acid* — which has powerful physiological actions that have been highly-touted in the popular press. (I discuss CLAs in more detail in my article "Conjugated Possibilities".) Note that CLA is *not* made by the plant itself. *CLA is only an intermediary molecule* created during the stepwise saturation process of linoleic acid in the rumen.

And CLA is only absorbed into the blood under rather specific conditions. When animals are fed a high-grain diet, the rumen environment forces the fat saturation process to proceed quickly from step to step, so intermediary compounds like CLA do not accumulate. But when animals consume their feed by grazing forages, the rumen environment favors a slight accumulation of CLA molecules. Some of these CLA molecules are then absorbed by the animal across the gut wall and are deposited in its fat — in its body fat and milkfat. This helps explain why grassfed animals contain higher levels of CLA than grain-fed animals.

Now for the practical issues of using fat in rations . . .

Oh my. I see that we are out of space. Again. Hmmm. Well . . . we'll complete this discussion next month. I promise.

First Published: November 2008

Author's Note: Fats are generally not a significant component of any ruminant diet, but they certainly garner a lot of publicity. For all of us, a little knowledge of the principles of fat chemistry will go a long way.

Feed Soaps

So . . . if fat is so high in digestible energy, and extra energy can benefit high-production diets and cold-weather diets, why don't we just dump a lot of fat into our rations and call it good? Well, because if we do, the results will not be good. Our animals would reduce their feed intake and possibly even go off feed. Production would go down, not up, and cold-stressed animals would feel colder than before. Let's explore why and what we can do about it.

Chemically, the basic unit of fat is the *fatty acid*. Although details vary between fatty acids, they all share the same basic structure: they are all composed of a single string of carbon atoms like individual links in a chain, with a methyl group ($-CH_4$) at one end of the chain and a carboyxl group (-COOH) at the other end of the chain. And since most feedstuff fatty acids contain a chain of 16–18 carbons or more, they are called *long-chain fatty acids*.

In nature, these long-chain fatty acids generally don't exist as freestanding molecules. They are usually linked to a glycerol backbone — glycerol is a simple 3-carbon alcohol — to form a molecule with three fatty acids called a *triglyceride* (tri = three. It logically follows that a molecule containing *two* fatty acids is called a *diglyceride*. And a molecule with only *one* fatty acid is called a *monoglyceride*. But a molecule with *no* fatty acids is *not* called a *nonglyceride*).

Most fats that we would add in a ration — such as vegetable oil or the fat in corn — occur naturally as triglycerides. Things get interesting, however, when that ration is consumed by a sheep or cow . . .

When a triglyceride enters the rumen, the molecule dissociates into its component parts: three separate free fatty acids and one molecule of glycerol. Rumen bacteria quickly ferment the glycerol for energy, but these bacteria can't ferment the fatty acids because they don't have the necessary biochemical machinery. Rumen bacteria can, however, make some changes to fatty acid molecules. Last month, I described how the bacteria convert unsaturated fatty acids into saturated fatty acids. But fat molecules have another effect in the rumen. Surprisingly, unsaturated free fatty acids are *slightly toxic* to certain types of rumen bacteria, and under some conditions, these fatty acids can have a profound effect on the rumen environment.

Let's cut to the chase: large amounts of unsaturated fatty acids in the rumen tend to suppress the methane-forming bacteria. These are the bacteria that produce methane gas as a byproduct of their fermentation of fiber. Translation: a high level of vegetable fats in a ration reduces the fermentation of fiber. Which means that fiber doesn't leave the rumen as quickly as previously — its retention time in the rumen increases — which means that the animal tends to feel "full." An animal feeling satiated doesn't want to eat more feed. Therefore, putting too much fat in a ration can trigger a nutritional paradox: *increasing* the energy value of a ration by adding fat can *reduce* feed intake, which will cause a *net reduction* in the amount of energy received by the animal.

The obvious question is how high is "too high?" Here's a rule of thumb: *avoid fat levels higher than 6–7% of the total ration.* Lower fat levels can provide a net gain of digestible energy to the animal, but fat levels above this threshold increase the risk of reduced feed intake. Individual feedstuffs, of course, can contain high levels of fat — the 6–7% limit only applies to the *total ration.* For example, pure vegetable oil contains, well, 100% fat. In practice, we can usually safely add 2–3% vegetable oil to most rations. And a popular byproduct of the ethanol industry — dried distillers grains (DDG) — contains approximately 10% fat on a DM basis. But more on DDG in a moment.

So are we always limited to feeding fats at levels below 7%? Well, there *is* a way around this limit. We can feed *soaps.* At least certain types of soaps. Read on.

We need to get specific. Some rumen bacteria are suppressed by high rumen levels of *free fatty acids,* particularly unsaturated fatty acids. The triglycerides that occur in feeds are safe, but when these triglycerides enter the rumen, they dissociate into those pesky free fatty acids. But if we can offer a feedstuff that combines free fatty acids with something that does not dissociate in the rumen . . . hmmm. That something could be calcium atoms or magnesium atoms. Attaching a calcium atom to the carboxyl end of certain fatty acids would create a stable, rumen-inert molecule. This molecule could then move safely through the rumen and deliver its fatty acid to the small intestine for absorption, without suppressing the fiber-digesting bacteria in the rumen. A molecule which contains a calcium (or magnesium) atom combined with a fatty acid is something we call a *soap.*

Of course, I'm not recommending that you run out and grind up 200 lb of your perfumed bathroom soap and dump that onto your hay. But some companies have indeed created ingeniously-designed commercial feed soaps. For example, one of the first commercial fat products was MegaLac® which is a combination of palm oil fatty acids and calcium. Animals with high energy requirements, like high-producing dairy cows, can nutritionally benefit from this type of feed additive, although the dairyman must carefully work out the cost/benefit numbers for his own situation. I suspect that these compounds can also provide extra energy for dairy sheep and goats, as well as productive ewes trying to raise triplets.

Let's get back to DDG, which is the byproduct of the ethanol industry. Essentially, the entire industrial process is designed to ferment corn into ethanol. This process uses up all the starch and leaves a residue called DDG that contains all the other grain nutrients in more enriched levels. Since corn grain naturally contains 3–4% fat, the DDG contains approximately 10% fat, although this level can vary slightly depending on the specific industrial process.

For years, research has demonstrated that DDG has more digestible energy than corn. This makes sense — and one reason is that it contains lots more fat. But there are limits. We cannot feed ruminants a ration of 100% DDG — that would exceed our threshold of 7% fat. A ration of 40–50% DDG may be more reasonable, depending on what else is in the ration. But now we have a general guideline for using a feedstuff that is relatively high in fat.

On a practical basis, of course, fats have other attributes. They can be used to reduce dust in rations, or to bind pellets, or to alter the flow characteristics of a custom feed mix, or increase palatability. Or decrease palatability if they become rancid. Each fat has its own quirks, but that's part of the art and experience of manufacturing feeds.

The bottom line is that we should always remember the upper limits of adding fat to ruminant diets. The rest . . . well, this month I'll refrain from trying to use the concepts of "soap" and "making a clean feed mix" in the same sentence.

First Published: December 2008

Author's Note: Adding fats to rations entails some knowledge, skill, and experience. But when circumstances permit it, we can use fats judiciously to refine our rations and gain an extra edge.

The Real Skinny on Fat

Search the Internet for the term "CLA", and you'll find some intriguing references: California Library Association, Caseous Lymphadenitis (a sheep disease), Computerized Laryngeal Analyzer, Conjugated Linoleic Acid. I know about the first two; I haven't a clue about the third one (and I'm not sure that I want to know). The fourth one, well . . . Conjugated Linoleic Acid is *the* hot topic in medicine and human nutrition. Also in animal nutrition. Why the big deal?

Because research is showing that CLAs may have profound effects on certain forms of cancer. And that natural CLAs can only come from the fat of *ruminants*. Particularly grazing ruminants, not grain-fed ruminants. Now *that* is something to chew the fat about.

But CLA is a type of fat. It's difficult to describe fats properly without using some technical terms, and fat terminology is, well, rather convoluted. To most people, the names of fats are befuddling concoctions of common names, numbers, codes, and technical terms. And the current marketplace hype doesn't make it any easier. So before we can intelligently discuss CLAs, we need to cover some of the basics about fat.

CLA is a specific type of fat called a *fatty acid*. Fatty acids are the main components of all animal fats (tallow, lard, milkfat) as well as plant fats (oils). On a molecular level, a fatty acid is simply a chain of carbon atoms, like a string of railroad cars, with lots of hydrogen atoms attached to those carbons. This chain is not symmetric: one end of a fatty acid chain always consists of a group of atoms called a *carboxyl group*, and the other end terminates as a *methyl group*, which is a carbon atom bonded to three hydrogens. Along the length of this chain, each carbon atom is linked to its neighbor carbon by either a single bond or, less frequently, a double bond.

Now for the names: scientists describe fatty acids by two main characteristics: (1) the number of carbon atoms in the chain, and (2) the number of double bonds. By convention, the standard fatty acid terminology uses two numbers separated by a colon. The first number is the number of carbon atoms; the second number is the number of double bonds. For example, a fatty acid that contains 18 carbon atoms and three double bonds would be labeled as C18:3. Most fatty acids in animal fats and plant fats contain 16–20 carbons. Therefore, typical fatty acids in animals would be identified as C16:0, C18:3, etc. (By the way, the most common natural fats are not odd — they usually contain an even number of carbon atoms, such as 14, 16, 18, etc.)

Of course, scientists are people too, and they don't like calling anything "C18:0" any more than you would. So they've labeled many fatty acids with common names like palmitic acid (C16:0), stearic acid (C18:0), oleic acid (C18:1), linoleic acid (C18:2), linolenic acid (C18:3), and arachidonic acid (C20:4). Other fatty acids have common names too, and we'll get to some of them later.

One term that nearly everyone has heard about is *saturation* (as in "saturated fats"). This really refers to the frequency of double bonds in the molecule. If a fatty acid contains no double bonds, it's called *saturated.* If a fatty acid contains *any* double bonds, it's called *unsaturated.* If the fatty acid contains two or more double bonds, it's called *polyunsaturated.*

Let's focus a little more on those double bonds — mainly where they are located in the chain and how they are structured. Scientists commonly mark the position of a double bond by the number of its *first* carbon atom, and they begin counting at the carboxyl end of the chain. For example, if the 9th and 10th carbons were linked together with a double bond, this double bond would be termed in the #9-position.

Also important is the precise geometrical structure of these double bonds. A double bond can occur in one of two very different arrangements — *cis* or *trans.* Imagine the following: think of a single bond as a universal joint — the two components in a universal joint can swivel easily in any direction. In contrast, think of a double bond as a stiff elbow joint — angular, fairly rigid, with each component pointing in a specific direction. In the world of fatty acids, the *cis* and *trans* double bonds are like two types of elbow joints. Because the *cis* form is slightly more stable energetically than the *trans* form, the *cis* form is much more common in nature. But both forms exist, as we will see, and the specific *cis-trans* combinations in fatty acids are very important indeed.

One particular arrangement of double bonds deserves special attention. Fatty acids sometimes contain two double bonds separated by one single bond. This three-bond arrangement is called *conjugated.* For example, if an 18-carbon fatty acid contains a double bond between carbons #9 and #10, then a single bond between carbons #10 and #11, and then another double bond between carbons #11 and #12, that molecule would be called a *conjugated fatty acid.* Because this example is a C18:2 fatty acid with the common name of linoleic acid, this particular molecule would be called — you guessed it — *conjugated linoleic acid (CLA).*

There we have it. We now understand enough about the fatty acid terminology to describe fats accurately and easily. So here are some basic statements about fats. Get comfortable and follow along:

All the long-chain fatty acids contain approximately the same amount of energy — 9 kcal/gram. This includes animal fats and plant fats, saturated fats and unsaturated fats. Plants typically contain lots of unsaturated fats like C18:1, C18:2, and C18:3. When sheep or cattle consume these unsaturated fats, their rumen microbes add hydrogen atoms to the double bonds and thus convert these fats into saturated fatty acids such as C18:0. Fat from hogs, however, may be quite different than fat from sheep and cattle. Since hogs don't have rumens, the fatty acids stored in their bodies reflect the unsaturated plant fats consumed in their diets. Therefore swine fat (*lard*) tends to be relatively unsaturated because the plant fatty acids are unsaturated. Since sheep and cattle have rumens, their body fats (*tallow*) are more saturated than swine fats or plant oils.

Here's a mind twizzler: In general, as the number of double bonds in a fatty acid *increases*, the melting point of that fat *decreases*. For example, polyunsaturated plant fats tend to be liquid at room temperature, while the more saturated animal fats tend to be solid. Think of corn oil versus butter.

Finally, scientists have recently focused a tremendous amount of research on one specific 18-carbon fatty acid — *cis*-9, *trans*-11 C18:2. Notice that its two double bonds are in a conjugated arrangement, and that one of these double bonds is a *cis*-9 bond and the other is a *trans*-11 bond. *This* is the CLA found in grass-fed ruminants that may have major medicinal effects. It even has a common name — *rumenic acid.*

More next time.

First Published: January 2000

Author's Note: Fatty acid terminology is actually quite common in today's world. Think of saturated versus unsaturated fats, or trans fats, or conjugated fats. Terms that pop up frequently in the popular press and, of course, on the Internet. But it all makes more sense once you understand the codes.

Conjugated Possibilities

In a 1935 biochemistry journal, the scientist R.G. Booth and his colleagues reported an intriguing item: they observed that when dairy cows were turned out to spring pasture after spending a winter in the barn, their milk contained elevated levels of conjugated fatty acids. In 1985, Michael Pariza of the University of Wisconsin published the startling observation that *something* in cooked ground beef reduced skin cancers in laboratory mice. These two reports, seemingly unrelated, were the beginnings of the current flood of scientific research about a family of fat compounds that may profoundly affect how we raise sheep, beef cattle, and dairy cattle and how we market their products. The compounds? *Conjugated linoleic acids (CLAs).*

Last month in my article "The Real Skinny on Fat," I covered some of the technical terms of fat chemistry. Now, the reward: let's use those terms to describe the details of CLAs. (You may want to find that article and review it. I'll wait.)

As you may recall, CLAs are subtypes of a fatty acid called *linoleic acid*, which is a common unsaturated plant fat (oil) containing eighteen carbons and two double bonds. Linoleic acid is affectionately known among chemists as C18:2. The two double bonds in CLAs are *conjugated*, which means that they are separated by one single bond. There are many types of CLAs. We are particularly interested in a CLA that contains double bonds in the 9th and 11th positions. But every double bond has two possible spatial configurations called *cis* and *trans*, so we must be very specific when we describe this CLA because its two double bonds can have a total of four possible configurations (the four = *cis-cis, cis-trans, trans-trans, trans-cis*), and these configurations may differ greatly in their physiological effects. Specifically, we are most interested in the CLA that is *cis*-9, *trans*-11 C18:2. This molecule has the common name *rumenic acid.*

In his landmark 1985 paper, Michael Pariza reported that an organic compound derived from grilled ground beef, when applied to the backs of mice, reduced the incidence of a certain type of skin tumor. We now know that the metabolically active substance in that meat was CLA or a collection of CLAs. Over the past few years, in scores of laboratories all over the world, scientists have observed similar results involving CLAs and cancers, especially mammary, skin, and prostate cancers. Even the National Research Council, a determinedly conservative organization, acknowledged in a 1996 report that CLAs can indeed inhibit cancer in experimental animals. But that's not all. There is tantalizing experimental evidence that some CLAs may affect more metabolic systems than just cancer. Scientists have broadened their investigations about CLAs to such diverse topics as body composition, blood clotting, immune function, cell growth, atherosclerosis, and the integrity of red blood cells. I even found one recent report on the effects of feeding CLAs to chickens to improve egg quality during refrigerated storage.

Most of these positive observations involve one specific compound: rumenic acid.

Not only is rumenic acid metabolically potent, but from the perspective of livestock producers, rumenic acid also has two critical characteristics. First, the main natural source of CLAs is the fat of ruminant animals (sheep, cattle, goats, etc.). Plants don't contain significant levels of CLAs, although they can contain linoleic acid. CLAs are *derived* from linoleic acid, and in nature, this conversion only occurs in the rumen.

Why? Well, here things get a bit technical: the shorthand description of linoleic acid is *cis*-9 *cis*-12 C18:2. (Note that natural linolenic acid is *not* a "conjugated" fatty acid because its double bonds are separated by *two* single bonds rather than one.) When ruminants graze forages, some linoleic acid enters the rumen along with the other plant material. The rumen bacteria don't digest this linoleic acid for themselves, but their enzymes do alter some of the chemical bonds. Through a series of complex reactions, the rumen bacteria shift one of the double bonds from the 12th position to the 11th position and also change the configuration of that bond from *cis* to *trans*. The result is a transition molecule, which is then usually quickly converted to something else. But under certain rumen conditions, some of these molecules, now proudly brandishing the name of *rumenic acid*, escape the rumen by washing downstream into the small intestine where they are absorbed across the gut wall. The animal then incorporates these rumenic acid molecules into its body fat and milk fat.

The second important characteristic about rumenic acid is that its levels are more than *twice as high* in animals grazing green forages than in animals fed grain or hay. CLA production is very sensitive to rumen conditions. Fresh green forage supplies enough dietary linoleic acid into the rumen and also supports the proper rumen environment for the production and absorption of rumenic acid. Different feeding regimes — such as high-grain diets or even dry hay — do not result in these rumen conditions, and therefore very little rumenic acid washes out of the rumen. The increase of milk CLAs that Booth observed in 1935 when he opened the barn doors and turned his dairy cattle onto young spring pasture was the direct result of changing the diet from winter hay to fresh grass.

Rumenic acid, however, leaves us in an intriguing situation. For years everyone has heard the mantra of reducing the intake of fats, especially of switching from saturated animal fats to polyunsaturated plant oils. Now we are faced with the real possibility that the fat of beef cattle and sheep and the milkfat in dairy products, under certain conditions, contain higher levels of a compound that may be valuable in human health and metabolism. And that those certain conditions mean *grazing* — a management system that is financially viable and environmentally friendly.

Of course, nearly all the medical research with CLAs has been conducted only on laboratory animals, not humans. And although most of us are not mice, rats, or chickens — something worth remembering when we hear the health claims touted by commercial vendors of CLA pills — the scientific evidence *is* becoming very impressive. Looking down the winding road of scientific investigations, I think there is a reasonable chance that future research will demonstrate that some of these CLA effects will also apply to humans.

The bottom line is that the phrase "grass-fed" may have an explosive marketing potential for our products — and it would be based on real science. But think of this — if we in North America are slow to capitalize on this emerging body of knowledge about the relationship of CLAs and grazing ruminants, others may be faster. For example, the farmers in New Zealand and Australia already raise nearly *all* their livestock on grass. And they are well aware of the value of marketing. It's something to ruminate about.

First Published: February 2000

Author's Note: We've come a long way. From 1935 to the early 21st century, from the careful recording of a curious milkfat anomaly to the very detailed metabolic reactions of a powerful transitory compound. CLAs represent a change in our knowledge, another layer of complexity in our understanding of nutrition and intermediary metabolism.

SECTION 2

Minerals

Please Pass the Salt

Salt, minerals . . . minerals, salt . . . Aren't these the same things? Everyone, it seems, has a question about minerals, like "why do some mixtures contain only 4% salt while others contain 90% salt?" or "why won't my animals eat it?" This topic is like a Byzantine market, with all sorts of ballyhoo and potions. Well, let's just dive in . . .

First, I'll try to clear up some jargon. In the feed store and in farmer/rancher parlance, the word "salt" means sodium chloride — simple white salt. One atom of sodium combined with one atom of chlorine. NaCl — the same stuff you put into soups. Actually, the rest of the chemical industry uses the word "salt" to mean nearly *any* combination of positive and negative charged ions, such as magnesium sulfate or potassium chloride or calcium carbonate. But for agriculture let's stick with simple sodium chloride.

The term "minerals," on the other hand, means a *mixture* of things: salt (NaCl) *and* other compounds containing elements like magnesium or potassium or iron or copper, usually combined in a lick block or in a bag of loose minerals. That's what you see when you walk into a feed store. But this farm definition would greatly displease a chemistry teacher, because in chemistry the word "mineral" generally applies to many types of inorganic substances, like magnesium or rock phosphate or borax or diamonds. And in certain areas of the country, some ranchers even call white salt their "mineral," which makes everything as clear as mud.

Here, I'll refer to the mixture of salt plus other minerals as a "trace mineral mix" (*TM mix*).

One main principle underlies all strategies for mineral nutrition: animals crave salt. That's why livestock lick those unappetizingly concrete-like mineral blocks — because they're trying to consume salt. Sheep and cattle don't necessarily *like* pure magnesium or selenium or potassium — in fact they usually avoid those chemicals. But feed companies know that animals want salt, so they add the less palatable trace minerals into a mixture with salt and call it "trace mineral salt" or a "trace mineral mix."

How much salt should be in a TM mix? Well . . . I've seen successful mixtures that contained anywhere from 4% salt up to 90% salt. There's nothing magical about the level — animals are very good at regulating their own intake of salt. They'll eat what they need. If a mixture contains less salt, animals will eat more of it. Companies design their trace mineral recipes to balance the intake of trace minerals with the anticipated intake of salt. But we live in an economic world, so here is a no-nonsense bottom line logic: feed companies that make mineral mixes only make money when they sell bags of mineral mixes. The more bags they sell, the more money they make. Animals will eat more mineral mix if the salt percentage is low. Therefore, how much salt should companies put into the mix? It's their judgment call.

However, when the TM mix is used to deliver drugs or certain elements — like selenium or Bovatec® or Aureomycin® — the consumption level is critical. Variations of intake will alter the dosage. Feed companies will sometimes try to stabilize mineral intake at a relatively high level by including palatable ingredients in the mix. How many times have you heard someone say, "My sheep really liked that mineral!" Look at the feed tag — it probably lists things like molasses or distillers grains or other flavor enhancers.

But what about the flip side? What about the times when animals don't seem to eat enough mineral mix?

Usually this occurs because we've offered a *second* source of salt at the same time. Sometimes we do this to "add some variation" and sometimes we do this without even realizing it — a theme with infinite permutations. I call this problem *Double Trouble.* For example, have you ever offered your animals a standard TM mix and also a bloat block *at the same time*? Or put white salt out *next to* the trace mineral mix? Or fed a commercial protein supplement or protein block that contains molasses, urea, and twenty-some other ingredients, *including salt*? Read the label. In this situation, animals will happily lick that protein block and ignore the nearby TM mix which contains selenium or a drug.

Or let's say that you buy hay for the winter, feed it, and observe that your animals don't seem to eat the mineral mix. Are you using *salted* hay? This is a technique that may be foreign to some folks, but salting hay is often used in the Pacific Northwest as well as other places around the country. When ranchers make hay and know that the bales are too wet, they sometimes sprinkle white salt on the bales to draw out the moisture. Salting hay has saved a lot of barns from burning down around here, but . . . when that hay is fed months later, the animals also receive a nice dose of white salt along with their hay. (Don't ever salt hay with a TM mix. Not only is the mixture more expensive than white salt, but the extra trace minerals can be toxic to animals).

And then there is the problem of selenium. Offering a free-choice TM mixture is a good, FDA-approved method of getting selenium into animals and preventing white muscle disease. Back in the 1970s, the FDA limited selenium levels in free-choice TM mixtures to 30 ppm for sheep and 20 ppm for cattle, based on extensive research on experimental farms. This looked okay on paper, but was a disaster in the field. Producers fed those TM mixtures and still saw lots of cases of white muscle disease. An entire generation of farmers concluded that feeding selenium in TM mixtures in fact didn't work, and they returned to their tried-and-true use of selenium injections.

The real culprit turned out to be the lower-than-expected intakes of the TM mixtures, which occurred either because of large variations in intake ("the mineral ran out, and we didn't get a chance to refill the feeder"), or the use of multiple sources of salt. Either way, if the TM mixture was the only source of selenium, the animals did not get enough of it.

In 1987, the FDA recognized this flaw and raised the selenium limits in TM mixtures to 90 ppm for sheep and 120 ppm for cattle. Reports of selenium deficiency dropped off immediately. The extra selenium provided a cushion that allowed for typical intake variations on a farm or ranch. These higher selenium limits are still in effect today.

So the next time you buy a trace mineral mix, read the feed tag . . . and add some spice to your day.

First Published: May 1998

Author's Note: The Double Trouble problem is really quite common, especially when animals are reared near barns. I am often called to operations where the producer is offering his prize animals a smorgasbord of goodies, some which contain salt. Just reading all the feed tags gives me the shivers. But, once we sort this out and find ways of providing only *one* source of NaCl, the mineral problems usually go away.

Double Trouble

We often go to great extremes to do the best for our animals. We build enormous barns, we manage topnotch pastures, we feed tasty grain supplements, we give medicines and vaccines, we stay up long nights teaching orphans how to drink from bottles, and we provide minerals and vitamins. And sometimes our well-meaning efforts get us into nutritional hot water, especially concerning minerals and vitamins. Actually more often than you would imagine. Perhaps the most common problem is something that I call *Double Trouble.*

The basic principle of minerals and vitamins is quite simple: livestock should get enough of them each day. Because only *some* minerals and vitamins are stored effectively in the body, our best strategy is to give them access to a mineral mixture daily — either free-choice or in prepackaged amounts — and assume that the mixture takes care of their needs. There is not enough room in this month's article to go into details about individual minerals, but there is a common belief among some folks that our livestock have some sort of internal "nutritional wisdom" about minerals and vitamins. Let me say this unambiguously: with one important exception, livestock *do not* have nutritional wisdom to choose minerals and vitamins properly. If offered a selection of trays, each containing an individual mineral, our animals would not select what they need in the correct amounts. In fact, since many mineral compounds are quite unpalatable, animals will stubbornly avoid those trays, even when they are dying of those mineral deficiencies.

The one clear exception is *white salt* — which means old-fashioned *sodium chloride.* In fact, "salt" is the official and legal feed tag name for sodium chloride. Livestock obviously relish salt. They seek it out when they need it, and they won't over-consume it to toxicity as long as they drink enough water to excrete the excess. The feed industry universally recognizes this feature, and companies mix salt with other less-palatable minerals (and vitamins and drugs), selling the product as a *Trace Mineral (TM) Mixture.* The percentage of salt in this mixture is not as critical as you might think. I've seen successful TM mixtures with salt levels ranging from 4% up to 96%. Each company formulates its own recipes, and each mineral recipe is carefully designed for a specific expected level of intake. In any case, the underlying concept for these TM mixtures is that *salt is the driving force of the mineral intake.*

And this creates the problem I call *Double Trouble.* If you offer animals two or more sources of salt (the "Double" in Double Trouble), what will happen to the intake of your main TM mixture? Either (a) mineral intake will go down or (b) mineral intake will become more variable — either over time (some days very little, some days very high) or within the flock or herd (individual animals responding differently to these choices). Or all of the above. You will have lost control of your mineral intake. And if you depend on that TM mixture to provide specific dosages of critical minerals like selenium or drugs like Bovatec® or other antibiotics, what will happen to the dosages of those ingredients? The dosages will decrease or become more variable — which will increase the risks of mineral deficiencies, reduce drug effectiveness, and increase microbial resistance to drugs.

Now let's talk about some Double Trouble scenarios that occur on farms and ranches in the real-world.

The most obvious scenario is to feed extra white salt. Yes, some people do this because they think (1) their TM mixture doesn't contain enough (or any) salt, or (2) to save money ("hey, my animals eat less of that expensive mineral when I offer white salt"), or (3) they have simply "heard" that white salt is a good thing. A variation of this scenario is to offer three or four or even more mineral mixtures — just to "make sure." One quick cure for this problem is to read the feed tag of the original TM mixture. If the feed tag specifically gives directions to feed white salt, then of course follow the directions. But if there are no such directions, then study the list of ingredients. If you see the term "salt," then you know that the feed company has already included sodium chloride in its original mixture, and you don't have to supply any additional salt. But if you feed extra white salt to reduce the intake of those minerals, you actually *dilute* the intake of the original mineral mixture, and you defeat the goals of the company nutritionists and expose your animals to all those health risks. And if you offer three or four different mineral mixtures at the same time, mineral nutrition *really* becomes a tangled mess.

Another Double Trouble scenario occurs when some folks routinely feed a grain or protein supplement. Sometimes this supplementation is necessary for production, sometimes not; but in either case, look at the feed tag of that supplement. Straight corn or oats or other grain don't contain salt, but a commercially-prepared grain mixture *may.* Grain mixtures are always *very* palatable. If animals eat one pound of a yummy supplement that contains salt, they're also consuming that extra salt. Again, how will this affect the consumption of your free-choice TM mixture? Many times I have visited a ranch where the owner proudly shows me how he feeds a little of this, a little of that, a scoop of this other stuff, and also a cupful of a special mix from that bag in the corner. Oh my.

Another variation: do you use a lick tank to provide extra energy or protein? A lick tank usually contains molasses and urea and perhaps some other ingredients or drugs. But you should read the label — does it also contain salt?

Here's something that may be a specialty of the Pacific Northwest — although I suspect it is used elsewhere — *salted hay*. On the west side of the Cascade Mountains we sometimes get a bit of rain during the haymaking season (that's a joke. Laugh. We *always* get rain during the haymaking season). Sometimes the square bales are too wet to stack safely in the barn, so we do this: after laying down a layer of damp hay bales in the barn, we generously sprinkle white salt on top of that layer. We do this for each layer of hay. Our hope, of course, is that the salt will draw enough moisture out of the bales to prevent the barn from exploding in flame. The existence of long-standing barns in the Pacific Northwest is kind of a backhanded proof that this technique works. But a secondary result of this technique is that the hay contains salt. When that hay is fed months or years later, folks may have long forgotten about the salt, but soon the animals begin suffering from unexpected mineral deficiencies. Double Trouble, again.

A variation of this scenario occurs when salted hay is *sold*. The unsuspecting buyer gets a truckload of hay, feeds it out, and unexpectedly runs into mineral problems like selenium deficiency. As a hay buyer, how can you protect yourself? Well, since hay does not usually come with a user guide, the most practical way is to monitor your animals' intake of the minerals. If mineral intake suddenly goes awry when a new source of hay is fed, you should become concerned. (The official recommendation, of course, would be to test that hay in a lab. That's fine and dandy — as long as your sampling technique is good enough that you can fully depend on the results.)

Here's another interesting Double Trouble scenario: the ocean. It's a big world out there, with more than 70% of it covered in water. Salt water. All along the coastline of North America, fields are exposed to ocean fog, spray, and wind. When I work near the ocean, I like to take a grab sample of the growing forage and analyze it for minerals, especially sodium. I generally expect to see background sodium levels lower than 0.20%, dry matter basis. A sodium level higher than 0.40% is a red flag. Salt in growing grass is still salt, and it's something to watch.

This Double Trouble theme has nearly endless variations, such as salt licks, bloat blocks, high-salt streams, etc. But once we identify the problem, what can we *do* about it? Some scenarios are quite easy to fix — for example, it's easy to stop feeding the extra bag of white salt. But what about those situations where we can't easily eliminate the second source of salt?

Let's return to the original concept of trace mineral mixtures. If mineral intake is driven by salt, and something interferes with the effectiveness of salt as an intake stimulus, then we should try changing the driving force of intake. Find an alternative TM mixture that contains other tasty ingredients, like flavor additives or molasses. This new mixture will probably also include salt, but the salt is only along for the ride, just like any other required nutrient. The real intake stimulant is something else. Something that can get you out of Double Trouble.

First Published: October 2007
Author's Note: How common is Double Trouble? Well, when I give nutrition workshops and describe this problem, I see that the people in the room become very quiet. They are listening carefully. They may not volunteer information or raise their hands to ask questions, but their intense silence and concentration tells me that I have hit home. I am describing a situation that they recognize from their own operations, and they are mulling over possible solutions as I speak.

For More than Strong Bones

Imagine that you are standing by the drop pen, checking the heavily-pregnant ewes. One mature ewe is down. Your first reaction — ketosis. But wait . . . although this ewe is obviously carrying twins, you see that she is in reasonable body condition. You're already feeding her some extra grain along with good alfalfa hay and minerals. You smell her breath. There is no telltale scent of acetone. Hmmm . . . ketosis may not be the problem after all. Well, you can drench her with propylene glycol anyway, but then you think — perhaps something else could be wrong here. Something like calcium.

What? A calcium problem from feeding alfalfa hay that contains more than 1% calcium? Oh yes, and for this ewe, the risk may be rather high. Ask any dairyman about alfalfa, mature animals, and milk fever. Calcium problems are quite common in the dairy world, and they probably occur more frequently in sheep and beef cattle than we know because the symptoms in pregnant ewes look very much like ketosis.

It may be helpful, however, if we first realign our thinking about calcium. This calcium problem is not a *deficiency* in the classic sense of a calcium deprivation in the diet. The diet may actually contain high levels of calcium. This problem is really a *dysfunction* due to low calcium levels in the blood.

This syndrome has many names — *milk fever*, *hypocalcemia*, or the formal name *parturient paresis*. Actually, hypocalcemia is probably more appropriate than milk fever, because (a) *hypo* means under, beneath, or less than (as in hypothermia, hypodermic, or hypothyroidism) and *calcemia* means calcium in the blood, (b) in sheep and beef cattle this syndrome occurs during late gestation before any milk production occurs, and 8 there is no fever. But we shouldn't discard the name *parturient paresis* — we can always use it when we want to impress our friends.

Whatever we choose to call it, hypocalcemia is quite complex. So before we discuss prevention and cure, we really need to learn some background about calcium physiology.

We have all heard the mantra in human nutrition that "calcium builds strong bones." This is certainly true — for humans and livestock — but it's also a major simplification. All mammals use calcium in other roles that are crucial and far-reaching. Inside the cells, calcium acts as a biochemical messenger, a powerful physiological switch that turns on or off various metabolic pathways. Calcium is also critical in blood clotting, enzyme activation, and the conduction of nerve impulses. The blood carries calcium to its cellular targets, and the body works very hard to maintain the blood levels of calcium within a tightly controlled range. The body does this by manipulating the activities of two major systems: (1) the intestinal tract which acts as the entry gate, and (2) the bone tissue which acts as the bank account.

The intestinal tract controls how much dietary calcium enters the blood from the gut. But this entry is not just a simple process of passive diffusion. Instead, the body controls calcium absorption with an active transport mechanism. Calcium atoms can only cross the gut wall if they are actively "ferried" by a specific calcium-binding protein called *calbindin*, which literally grabs the calcium ions on the gut side of the membrane and hauls them across the membrane to the blood side. And here's the trick: the body changes the *efficiency* of this molecular ferry — increasing or decreasing this efficiency — in response to changing calcium requirements and in response to the amount of calcium in the diet. For example, if dietary calcium levels are high, the ferry operates at a low efficiency (i.e., only a small percentage of the calcium atoms are ferried across).

The bone tissue, in contrast, is a vast skeletal bulk that can store and release calcium atoms. Most people think that bone is solid and unchangeable, but actually, bone tissue is in constant biochemical flux. Atoms move into and out of the bone matrix all the time, and the bone proteins and minerals are in constant turnover. (Very similar to money in our bank accounts.) The body alerts the bone about its calcium status through various hormones, including a derivative of Vitamin D. If circulating levels of calcium begin to drop, the bone tissue, like an automatic teller machine, releases some of its calcium into the blood. Mobilizing calcium out of the bone is called *resorption*. On the other hand, if circulating calcium levels are sufficiently high, the bone retains its calcium atoms in the skeleton.

We can think of these two systems as an elaborate two-phase set of controls. Bone mobilization is the quick-response system, capable of rapid day-to-day changes. The intestinal transport mechanism is the broader, slow-response system. Adjusting the intestinal system requires changing the amount of calcium-binding protein. Like any other industrial process, the manufacture of these proteins is a gearing-up process that takes time — as much as one or two weeks.

Under some conditions though, in spite of these control mechanisms, blood levels of calcium can still fall below a critical threshold. When that occurs, the animal suffers from the classic symptoms of hypocalcemia: muscular weakness, tremors, incoordination, restlessness, drowsiness, coma, and ultimately, death. And this all can transpire in a couple of days.

When are the risk periods for our livestock? In dairy cattle, the high risk period occurs soon after calving — the archetypal case of milk fever — when milk production rises quickly, pulling lots of calcium into the mammary glands. In sheep carrying twins or triplets, and occasionally in beef cows with a large, fast-growing calf in late term, the high risk period occurs during the last stages of gestation, when large amounts of calcium are moved into fetal bone and also into the colostrum. And just to make things interesting, sheep can also suffer hypocalcemia soon after lambing, especially if they have been stressed.

Not all animals are at risk. Age and stress are factors. Heifers and young ewes almost never suffer from hypocalcemia, and ewes with single lambs are safe. It's the older animals that are at risk. Apparently, some older individuals have a hard time mobilizing calcium quickly from their bones. And in marginal situations, an extra stress of stormy weather, or transportation, or even grafting an extra lamb onto a ewe can tip the scales.

So now we have the basic features of the hypocalcemia story: the physiology of calcium, its control mechanisms, the symptoms of the problem, and a sense of which animals are usually affected. Next time, we'll discuss the practical situations when things go wrong and what we can do about them.

First Published: April 2002

Author's Note: One thing we should remember about any gestation-related problem such as hypocalcemia is that the onset may cause an animal to go off feed, even a little. This, in turn, may precipitate the metabolic problem of ketosis. In the field, producers often rush to treat a sick animal for ketosis and then don't get the expected response. But if the primary problem is hypocalcemia, then *that* is what should be treated. Unfortunately, time is critical, so knowing the relationship between ketosis and hypocalcemia can possibly save quite a few animals over the years.

DCAD Tricks

In the field, we periodically encounter problems involving hypocalcemia. For example, I worked with one farm on the East Coast where the shepherd top-dressed his hay with limestone during late gestation because he recalled hearing "something" at a workshop about extra calcium before lambing. Another example: on a pasture-based ranch in western Oregon, I found that the early spring grass pastures contained calcium levels *lower* than phosphorus levels and also with very high levels of potassium, while at the same time, the rancher fed a trace mineral mixture that contained high levels of *both* phosphorus and potassium. To paraphrase an old song, the list goes on . . . The calcium situation in ewes and cows is indeed complex, but what can we do about it?

From an ambulatory perspective, we can play the roles of ER technicians when we encounter a downed animal: hypocalcemia shows up as a ewe or cow who is off-feed, wobbly, uncoordinated, listless, possibly showing tremors, and finally laying down and unable to rise. The good news is that we can usually fix this. The immediate problem is low blood calcium, so on the ER level, we add calcium back into the blood with an intravenous drip — usually in the form of calcium gluconate. This usually takes care of the symptoms. Veterinarians and farmers all have their favorite IV recipes which can also include dextrose, amino acids, and even some antibiotics. The glucose helps relieve any ketosis that may be occurring simultaneously. When it's successful, the IV procedure seems to work wonders — the animal often gets up and walks away, or it may even try to run us over.

But unless we want to make a lifetime hobby of giving IVs, our real goal is to *prevent* the problem, not just treat it. Our choices here are more complex, and they reflect the intricacies of calcium metabolism. Our basic strategy is to encourage the animal to increase its own blood calcium levels in a timely manner, and we have two basic methods for this.

The first method is one that dairymen have used for years to prevent milk fever: dairymen "trick" the animal's calcium transport system into responding to an artificially-induced deficiency just prior to calving. Since dairy cows require relatively low amounts of calcium during late gestation (compared to lactation), the normal situation is that the intestinal calcium transport system cruises along blissfully at low efficiency. This transport system has no need to absorb a high percentage of calcium because the background levels of dietary calcium are sufficient to meet the low requirements.

So, dairymen have traditionally capitalized on this situation by *reducing* calcium levels in the diet during the last 2–3 weeks before calving. This *tricks* the system by creating a calcium shortage, which stimulates hormones to cause the cow to increase the amount of calcium transport protein in the gut wall and also to accelerate the amount of calcium released into the blood from the bones. It takes a couple of weeks to make these metabolic adjustments, but then the cow calves and begins to produce milk. Although her calcium requirements then rise very quickly, the dairyman also immediately increases her dietary calcium levels. The pre-calving manipulations of the diet have already primed her biochemical mechanisms into high gear, so now those mechanisms can pour lots of calcium into the blood and prevent milk fever. Clever!

This technique is not perfect though, and one main drawback is that farmers can't always reduce dietary calcium low enough, especially in some forage-based diets. But dairy researchers recently discovered an alternative, more elegant approach. In the 1980s, Elliot Block from McGill University and other researchers observed that they could prevent milk fever by changing the proportions of certain minerals in the late gestation diet, even without altering calcium. From their empirical observations, they devised an ingenious equation called the *Dietary Cation-Anion Difference* (DCAD).

Don't let the terminology scare you. Think back to your high school chemistry. *Cations* are atoms with a positive electrical charge — these include minerals such as potassium and sodium. *Anions* are atoms with a negative electrical charge — these are other common minerals such as sulfur and chloride. To calculate the DCAD, you need to know the levels of these four minerals in the diet. But knowing these percentages is not quite enough, because the biochemical effects of each mineral must be adjusted for its atomic weight and its electrical charge. Therefore the DCAD equation also contains coefficients that correct for these two characteristics. Putting all this together, the researchers came up with this working equation:

$$DCAD = \begin{array}{l}(435 \text{ x sodium\%}) + (256 \text{ x potassium\%}) - \\ (282 \text{ x chlorine\%}) - (624 \text{ x sulfur\%})\end{array}$$

The DCAD result is expressed as *milliequivalents per kg of diet dry matter* (abbreviated as *mEq/kg*). Typical values range from +500 to –150.

All these numbers come to this: during the high-risk period for hypocalcemia, reducing the DCAD from a positive number to a negative number, or even reducing the DCAD from a high positive number to a much lower positive number, stimulates the cow to move more calcium from her bones into her blood and also encourages production of more calcium transport protein for increasing the intestinal absorption of calcium.

Whew! What planet did this come from? Yes, it *does* look like something out of science fiction book. In practice, however, many nutritionists now use DCAD values to guide their dietary formulations during high-risk periods. For example, the DCAD tells us to avoid diets high in potassium and sodium during late gestation.

On confinement dairy farms, where cows conveniently receive their feed as a *Total Mixed Ration (TMR)*, where all the ingredients are mixed together including minerals and vitamins, manipulating mineral levels is relatively simple. Some nutritionists will reduce the potassium and sodium levels of the late gestation diet and also add *anionic salts* to it. Anionic salts are minerals containing chlorine or sulfur without sodium or potassium — such as magnesium chloride or calcium sulfate. Look at the DCAD equation: anionic salts make the result more negative. Although anionic salts are generally unpalatable, in a TMR they can be camouflaged by tasty ingredients like molasses.

Pastures, however, pose a real challenge with high-risk animals — because forages can have particularly inconvenient levels of these four minerals. High quality forages may contain high levels of potassium and even sodium. A trace mineral mixture containing sodium chloride can arithmetically push the DCAD upwards because of its sodium. Changing the mineral levels of growing forages is cumbersome, if not impossible, and trying to add unpalatable anionic salts to a free-choice mineral mix may simply reduce mineral intake. In practice, manipulating the DCAD effectively for these animals may not even be feasible.

Reference books say that hypocalcemia in ewes is relatively rare. I don't agree. I've seen it in the field too often to be comfortable with the word "rare," especially when its symptoms resemble those of ketosis. In fact, pregnant ewes suffering from hypocalcemia go off feed which then can *cause* ketosis.

But unfortunately, there's a kicker — *these prevention techniques have been researched for dairy cattle, not for sheep.* This is like the story of some medical doctors who are trying to develop a new drug treatment regime for their patient. They know that their drugs have been thoroughly tested on rats and rabbits, so their task is to determine if their patient more resembles a rabbit or a rat.

Nonetheless, we still must deal with the problem in the field, and ewe biochemistry is similar to cow biochemistry. We can use the concepts from bovine research as our guides, add a dash of good judgment, and keep checking to see if our ewes begin to resemble rats or rabbits.

First Published: May 2002

Author's Note: The DCAD equation is a real breakthrough in ruminant nutrition — the result of careful observations and good science. We still need to see if and how it can be effective with other animals, especially sheep and goats. This is a good research project waiting to be explored.

Too Little, Too Much, Early Spring

Early spring — the first full warmth of sunshine, trees adorned with bright green leaves, the iridescent green of early grass pastures. But one morning you walk outside and notice your cows or ewes trembling, walking unsteadily. Suddenly one falls down with an awful tetany, frothing at the mouth, convulsing, and pawing the air. You frantically call your veterinarian. She comes right away, inserts an IV tube and, an hour later, the animal is back on its feet, looking rather pert and kind of surprised by the ordeal. "Oh yes," you think, "early spring . . . magnesium tetany. Where is that nutrition book?"

This is a problem with lots of names — *magnesium tetany, grass staggers, grass tetany, hypomagnesemia, winter tetany*, even *wheat pasture poisoning.* Whatever we call it, it's the result of low blood levels of magnesium, and it involves forages. Since magnesium is tightly associated with nerve impulses, affected animals show dramatic neural symptoms like agitation, incoordination, and convulsions.

But magnesium tetany is *not* simply a lack of magnesium; it's a complex syndrome that is still not completely understood. Treating the problem, however, is relatively straightforward: we supplement magnesium to our livestock. For animals showing overt tetany symptoms, veterinarians intravenously dump a huge load of magnesium into the blood (usually as calcium-magnesium gluconate), which usually gives spectacular results — blood magnesium rises, and the animal quickly gets back on its feet and regains coordination. For the rest of the herd or flock, we add extra magnesium to the diet for the duration of the high-risk period, either as magnesium oxide in the trace mineralized mixture or as a topdress on the hay or grain.

Magnesium tetany occurs sporadically. Some years seem worse than others, and some animals seem more susceptible than others. We usually view this problem in terms of risk, which is influenced by many factors. The most important is the level of magnesium in the forage. A simple rule of thumb for assessing forages: Magnesium levels above 0.18% are generally safe. Levels between 0.12–0.18% are medium risk. Levels below 0.12% are high risk.

But magnesium tetany is a tangled web that involves other risk factors. For example, we know that tetany usually occurs in the early spring on lush, young pasture. More cases coincide with cool, rainy weather that often occurs during the spring. But magnesium tetany can also occur at other times of the year, which is confusing. We also know tetany primarily affects mature animals rather than young animals, especially during late gestation or early lactation, probably because older animals are slower than juveniles to mobilize extra magnesium from their bones to supplement falling blood levels. And we also know that other nutrients can influence magnesium uptake and metabolism, such as potassium, calcium, and even nitrogen.

Let's look at potassium, both in the forage and in the soil. A high forage level of potassium can cause a double whammy: (1) it lowers magnesium levels in the forage by reducing magnesium absorption from the soil, and (2) it depresses blood magnesium in animals by interfering with magnesium absorption across the intestinal tract.

With forage potassium, how high is *high*? Well, I get concerned about forage potassium levels above 3.0% (dry matter basis). Grasses can be particularly guilty of this problem. Grasses, being grasses, love potassium in a way that some people seem to love chocolate — they'll eagerly consume more than they need. Although grasses require potassium for growth, if the soil contains excess potassium, grasses will accumulate potassium above their requirements without showing additional yield. In grasses, this phenomenon is called *luxury consumption* (in humans, it can still be called luxury consumption, but the effects of excess chocolate are definitely not agronomic). Legumes are not guilty of this sin, and thus potassium levels in clovers and alfalfa are never high enough to interfere with magnesium.

One common recommendation for reducing the risk of magnesium tetany is to add legumes to the pasture. This is a nice theory, but it's generally not very practical. Remember that tetany usually appears in animals grazing the earliest spring growth — when soil temperatures are still relatively low. Grasses grow at lower temperatures than legumes, so this early spring growth is — what? — grass. Any legumes in that pasture haven't yet begun to grow. An interesting conundrum.

So why is high forage potassium an issue, when so many farms and ranches struggle to improve their low soil levels by including potassium in fertilizer? Because in reality, some fields may actually contain high levels of potassium, either because those fields overlay high-potassium subsoils, or they have not been harvested for hay, or they've received lots of waste sludge from confinement dairy or hog operations. In general, soil potassium levels of 150–200 ppm will support good forage yields. But I've seen soil tests with potassium levels higher than 600 ppm, which is *very* high. I would want to check the potassium and magnesium of any forages growing on these fields, especially in the early spring.

Forage calcium also influences magnesium tetany. Experience has shown that tetany is influenced by a three-way interaction between forage calcium, magnesium, and potassium. Relatively low levels of calcium, combined with high levels of potassium seem to cause tetany problems, at least in forages with marginal levels of magnesium. But in assessing risk, most of us aren't comfortable with the phrase "*seems to,*" so agronomists have derived a handy formula called the *Tetany Ratio*, which combines the levels of these three minerals into a single number. A ratio above 2.2 indicates a high-risk situation.

The Tetany Ratio gives us a numerical tool to assess risk. It's particularly useful for marginal forages — forages with magnesium levels between 0.12–0.18% (although I would apply it to any forage with less than 0.22% magnesium). We don't need this ratio for the extreme magnesium values. Forages with less than 0.12% are clearly high-risk situations, and I would consider forages with more than 0.22% magnesium safe.

So what exactly is the Tetany Ratio formula? Oops, there's a problem here. Quite a few websites and printed documents do list a formula to determine the Tetany Ratio, *but unfortunately their formula is wrong because it is based on simple, uncorrected percentages. Using raw percentages gives misleadingly high values.* Try this yourself: create a simple spreadsheet to make the calculations easier and plug in a few values. You'll quickly see that even safe forages will give ratios higher than 3.0, which suggests a high-risk situation, even though the reality is a no-risk situation. So what is wrong?

Because the *real* formula for the Tetany Ratio is based on something called an "*equivalence basis*" which most of those published formulas ignore. In chemistry, "electrical equivalents" is a technical way of saying that the raw, simple percentages must be corrected for each atom's atomic weight and electrical charge. Each of the three elements has a different atomic weight and a different electrical charge. In other words, the formula must contain some correction factors. Here is the correct formula for the Tetany Ratio:

$$Tetany\ Ratio = \frac{(256 \text{ x potassium\%})}{(499 \text{ x calcium\%}) + (823 \text{ x magnesium\%})}$$

When *this* formula is applied to marginal forages, results greater than 2.2 indeed represent a good estimation of increased risk. In other words, the results coincide with reality (which is rather comforting).

Why has an incorrect formula been so widely distributed? Perhaps this is an example of people repeating each other without checking the results. Publishing something doesn't necessarily make it true or accurate. Be careful. When something doesn't seem right, it's prudent to check the original sources.

There are more theories and management implications about magnesium tetany, but we need more space. Let's talk about them next month.

First Published: March 2003
Author's Note: The confusion over the correct formula for the Tetany Ratio is unfortunate and causes problems. A number of official university documents simply include the wrong, uncorrected formula. I've worked with ranchers who were in a near-panic state after plugging their assay results into the incorrect formula. They were afraid to graze their spring pastures or were concerned about selling their hay. When we applied the correct formula, their forages looked a lot safer, and their personal stress levels declined immensely.

Magnesium Matters

We're talking about magnesium, tetany, and other magnesium things that matter . . .

Let's imagine ourselves on a farm during a high-risk season. Tetany has occurred here in the past, and our intravenous magnesium equipment is on-hand. We watch our animals carefully. Sure enough, we soon see the first animal with tremors and incoordination. But wait . . . let's hold off reaching for those IV tubes; we shouldn't *automatically* assume that magnesium is the problem, not just yet.

First, we should rule out other metabolic problems that cause incoordination, like ketosis and milk fever. Remember that magnesium tetany often occurs during late pregnancy and early lactation — prime periods as well for these other syndromes. Also we shouldn't discount the potential for polioencephalomalacia caused by thiamine deficiency or excess sulfur, especially if animals are grazing *Brassica* forages like turnips or rape. Also we should consider some *infectious diseases* that can cause incoordination like rabies, tetanus, and listeriosis (circling disease). Especially rabies — missing *that* diagnosis can ruin your entire week.

A word about confusing terminology: Be careful when reading articles in magazines or documents from the Internet. Terminology is a funny animal, particularly when others use it. Not every tetany relates to magnesium. The term *tetany* simply refers to muscle spasms or convulsive seizures, and the word *staggers* is a more dramatic portrayal of it. Lots of veterinary names contain the words "tetany" or "staggers," like *phalaris staggers* (caused by tryptamine alkaloids intrinsic to certain grasses), *ryegrass staggers* (caused by endophyte alkaloids in perennial ryegrass), *transport tetany* (probably caused by stress and lack of calcium), and even *gid* (also called Staggers, caused by a dog tapeworm infection of the sheep spinal cord). While these make interesting reading, none involve magnesium. It's kind of like the word *football.* In North

America, *we* know what football means, but to the rest of the world, football refers to a different game where people kick the ball with their feet — *soccer.*

But whatever we call this nutritional problem with magnesium, our best strategy is to prevent it. Magnesium tetany is more fun to read about than to treat.

The easiest prevention method is to add extra magnesium to the diet during the high-risk period, overwhelming the system with enough magnesium atoms to overcome any deficiency in the forage or interference from excess potassium. This is a sledgehammer approach, to be sure, but it generally works.

In practice, you simply replace your standard trace mineral mixture with one containing a high level of magnesium. Feedstores often sell "hi-mag" mineral mixtures, either as blocks or as loose minerals. Ideally, we'd like to see magnesium levels above 12% in these mixtures, but usually we're lucky to find levels above 7%. Some feed manufacturers may be reluctant to add much magnesium to their products because many magnesium compounds are unpalatable and the palatable ones that fix the problem are expensive. So read feed tags and compare products. Once you locate a good mineral mixture, offer it as the sole source of minerals during the tetany period — 4–6 weeks — and then switch back to your regular mineral mix.

If you raise sheep, however, what about copper? Although magnesium and copper are not directly related nutritionally, some commercial high-magnesium products may also contain high levels of copper, particularly if these products have been formulated for beef cattle. While that's okay for cattlemen, shepherds should be careful. The risk of copper toxicity depends on many factors including the total ration levels of copper and associated minerals (molybdenum, sulfur, iron), as well as the length of time this high-mag mineral mixture is offered. Sometimes there are no clear answers — just good judgments. But I digress.

An alternative strategy for supplementing magnesium is to mix it directly into a loose mineral mixture. You can usually buy magnesium oxide, which is unpalatable but readily available, and mix it with your current minerals together with dicalcium phosphate (dical) and corn meal or soybean meal, all in equal parts (i.e., in a ratio of 1:1:1:1). This strategy *may* work, but then again, if animals don't like it, they may not consume enough to avoid magnesium tetany, and they may also suffer from other deficiencies like selenium. Always monitor mineral intake and make changes quickly if it goes down — like adding more corn meal to improve palatability

Another technique for supplying extra magnesium is to put it in the drinking water. This is a good technique if you have a piped water supply, but it's not practical if your animals get their water from creeks or dew. Mineral solubility may also be an issue. Magnesium oxide is relatively insoluble, so you'll need to find a soluble form of magnesium, such as chloride, acetate, or sulfate. This shouldn't be too difficult, however, because magnesium sulfate is also known as *epson salts*.

And there are other clever options for supplementing magnesium as well, like mixing magnesium oxide directly into grain supplements, or adding magnesium acetate to molasses in a free-choice lick. A long-term strategy to increase forage magnesium levels would be to include magnesium in the fertilizer or lime, as in dolomitic limestone.

What about changing the *other* minerals in the ration? Last month, I discussed the *Tetany Ratio* and wrote that values above 2.2 indicate high-risk situations. So how can we lower this ratio? The arithmetically-inclined among us would see that we have only two computational options: either decrease the numerator (the top number, potassium) or increase the denominator (the bottom number, which is the sum of values for magnesium and calcium).

Decreasing the numerator means reducing the amount of potassium in the forage. This is not generally possible unless we change forages. Let's assume that for this year, we must live with our current forage, especially if it's our early-growth grass pasture. *Next* year we can try to change our forage and fertilizer. So let's look at the other option.

Increasing the denominator means increasing the amount of magnesium or calcium in the diet, or both. We've already discussed magnesium, but what about supplementing calcium?

Well, this option illustrates the difference between arithmetic and nutrition. Adding calcium to the diet would indeed arithmetically reduce the Tetany Ratio, and the result would be flooding the system with a large number of calcium atoms. But this has major nutritional ramifications. Too much extra calcium could interfere with magnesium absorption from the gut, which is exactly what we *don't* want. Also, extra calcium can reduce the absorption of other minerals, such as zinc, iron, and iodine, and thus create unexpected deficiencies of *those* minerals. And extra calcium at the wrong time could potentially increase the risk of milk fever in pregnant and lactating ewes and cows. So instead of falling down with magnesium tetany, animals would fall down with hypocalcemia. Not exactly an improvement.

Perhaps the option of providing extra calcium to prevent magnesium tetany is not such a great idea. Better to supplement extra magnesium in the diet now and consider ways of increasing forage levels in the future. Then after this year's tetany season, reflect on what happened, what worked, and what didn't. As every football fan knows, there's always next year.

First Published: April 2003

Author's Note: Magnesium tetany is indeed a frightening syndrome, but always keep in mind the differential diagnoses: look at the whole picture on the farm or ranch. The problem might still be magnesium tetany, but at least you'll be more confident with your actions.

Odiferous Consequences

Shepherds and cattlemen often talk about polio. Well, there's polio and then there's polio. And the second one has many causes. Let's talk about the differences.

The first polio is the human disease — *poliomyelitis.* This is the infectious human disease caused by a virus. This poliovirus, which spreads through contaminated water such as in public swimming pools, destroys neurons in the brainstem and causes muscle wasting, paralysis, and death. In the 1950s, Jonas Salk and Albert Sabin developed two wonderful vaccines that virtually eliminated poliomyelitis in the United States. This polio has nothing to do with animals.

The second polio is the animal disease — *polioencephalomalacia* (PEM). This is a neural disorder in sheep, cattle, horses, and other livestock, particularly in young stock. Animals with PEM show depression, incoordination, circling, blindness, star-gazing (*opisthotonos*), convulsions, and ultimately death. It's widely known that animals suffering from PEM sometimes respond quite dramatically to an injection of thiamine, and many farmers keep a vial of thiamine in their refrigerators for this purpose. Although ruminants and horses usually make enough of their own thiamine in the rumen or large intestine, toxins can destroy this thiamine before it's absorbed across the gut wall. Certain toxic plants, especially horsetail and bracken fern, contain the enzyme *thiaminase* which effectively destroys thiamine, and if livestock, especially horses, consume hay containing large amounts of these plants, they can suffer PEM due to a thiamine deficiency. This type of PEM generally responds to injectable thiamine.

But here's a paradox: while lack of thiamine can cause PEM, all PEM is *not* caused by lack of thiamine. Not all cases of PEM respond to thiamine injections. For example, PEM has occurred in feedlots where the high-grain diets did not contain those toxic plants, and it has also occurred in highly fertilized, well-groomed pastures that contain no toxic weeds. Indeed, cattle and sheep suffering from PEM in these situations usually show perfectly normal tissue levels of thiamine and don't respond to thiamine injections. So what gives?

Sulfur.

Actually, other things besides sulfur can also cause PEM, such as acute lead poisoning and sodium toxicosis due to water deprivation, but those situations are unusual. Let's concentrate on the far more common cause: sulfur.

Animals and plants universally require sulfur as a nutrient because it is an integral part of two essential amino acids, methionine and cysteine. All proteins contain sulfur. Nearly all feedstuffs contain some sulfur. We often fertilize pastures with the elemental form of sulfur or gypsum (calcium sulfate) or blended fertilizers. (For example, the fourth number on a fertilizer tag is the percentage of sulfur in the blend, as in 16-16-16-6.) Livestock need at least 0.14% sulfur in their diets, although this percentage is influenced by the amount and form of nitrogen in the diet. In sheep, for example, wool contains proteins high in sulfur, and wool growth can be limited by lack of it. Nutritionists are generally happy when they see dietary sulfur levels similar to phosphorus levels, namely between 0.15% and 0.28%. But a sulfur level of 0.35% is rather high, and nutritionists get very nervous when levels rise above 0.40%, which reference books usually list as the maximum tolerable level.

Once consumed by a ruminant, sulfur is dumped into the rumen and undergoes some fundamental chemical changes. Healthy rumens contain little or no free oxygen, but they contain lots of excess hydrogen atoms (where do you think that rumen methane comes from?). Rumen bacteria and protozoa are very good at obtaining oxygen from any compound in the rumen. Sulfur in feedstuffs is often in the form of sulfates or sulfites, both of which contain oxygen. Once sulfur compounds enter the rumen, bacteria strip those oxygen atoms and replace them with hydrogen atoms. And this process is accelerated in rumens of grain-fed animals, because feeding grain lowers the rumen pH — which means increasing the amount of hydrogen ions.

The resulting compound is a gas called *hydrogen sulfide* (H_2S). Under normal conditions, very little hydrogen sulfide accumulates in the rumen because most sulfide molecules are quickly assimilated into other compounds. But if the rumen contains lots of sulfur, especially in the form of sulfate, or if the rumen pH is low (which is common in grain-fed animals), hydrogen sulfide is formed in higher amounts, and this can be a problem because hydrogen sulfide is toxic.

You are probably familiar with hydrogen sulfide, because it has a distinctive odor. *Very* distinctive. Think of rotten eggs. The term "stench" is a mild understatement. Hydrogen sulfide is also affectionately known as *manure gas*.

Sulfur toxicity is nothing new. We already know that very high levels of sulfur can cause acute toxicity symptoms such as reduced intake, slow growth, labored breathing, scours, and even death. But in this article we're not interested in those acute symptoms. We're really interested in the more subtle relationship between sulfur and PEM, which seems to occur at sulfur levels not quite high enough to cause acute toxicity. Because researchers and veterinarians have noted that in situations where PEM did not respond to thiamine treatment, the animals usually had consumed relatively large amounts of sulfur.

Back to hydrogen sulfide. Since hydrogen sulfide is a gas, it initially accumulates in the gas cap of the rumen. Sheep and cattle routinely get rid of excess gas by *eructating* — a fancy word for belching. When ruminants eructate, some of that gas is actually inhaled back through the lungs (a rather sloppy design, if you ask me), and therefore some hydrogen sulfide can easily be absorbed through the pulmonary tissue. Many scientific studies have shown that hydrogen sulfide can cause the neurological symptoms of PEM. And these studies have also detected the characteristic odor of hydrogen sulfide in the gases eructated by the animals.

So . . . here's the bottom line. We should be alert for excessive sulfur in highly-digestible diets. We're not looking for acutely toxic sulfur levels, just levels high enough to be associated with PEM. And we should consider *all* parts of the diet. Some common feed ingredients can contain relatively high levels of sulfur, like molasses, rapeseed meal, and also certain byproducts of the grain milling industry, like dried distillers grains and corn gluten meal. Brassica plants (turnips, rape, kale) may contain high levels of sulfur. Pastures recently fertilized with sulfur, especially if there has been no rain, can have high levels of residual sulfur coating the leaves. And what about water? High-sulfur water with sulfate levels over 1,000 ppm can be quite common in some areas. Remember that a rumen's sulfur load is the *sum* of all these sources. Finally, low rumen pH levels — which are associated with feeding grain or even with very lush, young pasture — encourages the growth of the types of bacteria which produce hydrogen sulfide.

I suppose that you can monitor this situation by checking animals for bad breath. If you smell the odor of rotten eggs, you can infer that the sulfur level in the diet is, um, rotten. If, on the other hand, you are not familiar with the odor of rotten eggs, then leave some eggs out in the hot sun for a week or so. Just tell your family that it was a homework assignment.

First Published: January 2002

Author's Note: Is it *thiamine* or *thiamin*? To e or not to e. Ah, a question for the ages. One could spend a pleasant afternoon scanning through reference books and Internet sites, compiling lists of which ones use the long version and which ones use the short version. Some, like Webster's (very large) Third New International Dictionary, list both. I've chosen to use the long version. Perhaps I just think the extra "e" represents an unconscious sophistication. Some people will say that I am full of water, that it's obvious that I should have used the shorter version. Life is like that.

Copper Beneath the Surface

Let's talk about copper.

Specifically, let's focus on two knotty issues: how do other minerals interact with copper in the feed, and why are blood levels of copper sometimes so misleading?

First, some well-known facts: animals need only small amounts of copper each day. The National Research Council recommends that sheep need 7–11 ppm in their diets and cattle need 4–10 ppm. Copper is utilized by many metabolic systems in the body, including the production of hemoglobin. Excess copper is stored in the liver and eventually excreted in the bile. Most producers have heard that copper absorption is somehow affected by at least two other minerals — molybdenum and sulfur.

Molybdenum is more important than it appears. Molybdenum levels in grains and forages can vary from near zero to more than 3 ppm, and it essentially reflects the soil type. In general, molybdenum is much more available from soils with a high pH (greater than 7.0) than from soils with a low pH (less than 6.0). The critical fact is that, in the gut, molybdenum combines with the copper to form copper molybdates, which are not very soluble and will flow out with the manure. Copper atoms tied up in these molybdates cannot be absorbed into the blood. Sulfur plays a similar role: sulfur combines with molybdenum to form thiomolybdate, which can also form insoluble complexes with copper that will reduce its availability. Thus, high levels of molybdenum (over 3 ppm), particularly combined with high levels of sulfur, can reduce copper availability so much that an animal can actually suffer from copper deficiency.

But the opposite is also true: if molybdenum levels are very low (less than 1 ppm), no insoluble copper-molybdenum compounds will form. Therefore, copper will be absorbed without much hindrance . . . sometimes more than we want. In this situation, even normal copper levels in the feed can provide more copper to the animal than expected.

Historically, livestock producers have avoided copper deficiency by including copper in the mineral mixture. Copper deficiency causes the classic symptoms of "swayback," "steely wool," and a band of white wool in a colored fleece, as well as anemia and spontaneous bone fractures. But our issue here is not copper deficiency, but copper *toxicity*, which only recently has been recognized as a widespread problem. Sheep seem to accumulate copper more easily than other livestock, and over time sheep can amass enough copper in their livers to cause toxicity symptoms. In many areas of the country, particularly those with low levels of molybdenum in the feeds, chronic copper toxicity shows up as a sudden, spectacular crisis of jaundice and death.

Once copper atoms are absorbed across the intestinal wall, they are transported to the liver, which filters them from the blood before they can reach the rest of the body. Copper will remain in the liver (and to a lesser extent, in the kidneys) until it's either needed for hemoglobin synthesis or excreted through the bile. But in sheep, this excretion process is relatively slow — usually slower than the speed of entry — and over time copper tends to accumulate in the liver.

But copper atoms are very active chemically, which is a fancy way of saying that they can be toxic if they come into direct, uncontrolled contact with cellular molecules. Therefore, animals have developed a remarkable system for storing copper safely while protecting their own cells. This system is a protein called *metallothionein*. Metallothionein is a chelator (pronounced "kee-later") which grabs onto loose copper atoms inside liver cells and tightly enfolds them within a carefully-designed coil of amino acids — like a football player cradling a football. Each molecule can hold many copper atoms. Metallothionein is also *inducible*, which means that when the available supply of molecules is used up, the liver will make more of them. And metallothionein is fairly easy-going about its choices — it will also grab onto other minerals like zinc and cadmium. For all practical purposes, metallothionein is the liver's storage sink for toxic heavy metals.

Meanwhile, as copper accumulates in the liver, what do you think happens to copper levels in the blood? Nothing. And that is exactly what is supposed to happen — *because the copper filtering and storage systems in the liver are working smoothly*. Therefore, while copper levels in the liver rise slowly and insidiously, blood copper levels remain low — with no hint of any problem.

But even the metallothionein system has its limits. When copper levels in the liver eventually rise too high (above 300 ppm on a wet weight basis for sheep, although there is lots of variation among animals), things begin to break down . . .

Liver cells begin to disintegrate. Initially only a few burst, but as free copper spreads throughout the liver, cell destruction accelerates with frightening speed. Broken cells spill their contents into the blood, and these contents include copper as well as various enzymes that under normal conditions rarely show up in the blood.

Experienced veterinarians will tell you that an early sign of copper toxicity — even before plasma copper levels begin to rise — is the increased activities of two enzymes in the serum: GOT *(glutamic oxaloacetic transaminase)* and *lactic dehydrogenase.* These are liver cell enzymes, but their appearance in the serum indicates that, although the precise cause is not yet apparent, liver cells are being destroyed.

Finally, within a short period of time, vast numbers of liver cells die, all spilling their copper into the blood. Copper levels in the plasma rise suddenly and dramatically. The sheep goes down with the classic symptom of copper poisoning: the *hemolytic crisis.*

Unleashed copper is toxic to most cells, especially red blood cells. Copper-poisoned red blood cells rupture and release their hemoglobin directly into the blood. The red hemoglobin flows everywhere — into urine, mucous membranes, etc. — causing symptoms like *hemoglobinuria* (hemoglobin in the urine), anemia, and jaundice.

The hemolytic crisis is a physiological catastrophe. The outlook for that animal is grim.

We must ask ourselves: why did *this* animal show symptoms? *And what about the other sheep in the flock?* Remember that if we take blood samples from these other sheep, *their* copper levels may still be low. But do those low values assure us that all is well? No.

Therefore, once one animal shows the symptoms of copper toxicity, you know that your flock is on thin ice. Get them off that ice. Change the feed and management. Carefully.

First Published: October 1995

Author's Note: Chronic copper toxicity is a real problem in the sheep industry. Sources of excess copper can actually be quite plentiful on a farm. Read some of my other articles for these exciting details. And if you look on the Internet, you'll find long discussions among shepherds that will convince you that copper is the worst thing since the Huns attacked Rome. But keep the facts in perspective. Copper can indeed be an issue, either as a deficiency or as a toxicity, but copper is also a required nutrient, and a little knowledge here has the potential of being very helpful.

Copper Redux

If you own sheep . . . you know about copper. Every shepherd, it seems, can tell horror stories about unfortunate producers who've lost sheep to chronic copper toxicity (CCT). So the word is out, "Toxic! Toxic! Avoid Copper At All Costs!" Simple and easy to repeat, like a mantra, but unfortunately, it's not quite true, nor is it easily followed. Yes, copper is readily toxic to sheep, but it is also a *required nutrient* for them, and the difference between the two levels is rather small. And new research may make it harder to keep sheep away from high levels of it. Oh, the tangled web it weaves!

This topic is fairly involved, so let's all get on the same page. Most livestock species need 6–11 ppm copper in their total diet. Copper deficiencies show up as neurological problems (*swayback* in newborn lambs), loss of pigment in hair and wool, reduced immune function, reduced fertility, spontaneous bone fractures, etc. *High levels* of certain other minerals like molybdenum, sulfur, or iron will interfere with copper absorption across the gut wall. Conversely, *very low levels* of these minerals will potentiate copper absorption — i.e., make copper *more* available to an animal. And some geographical regions are actually *copper deficient*, either because the forage levels of copper are too low, or because the levels of molybdenum, sulfur, and/or iron are too high. Therefore, any feed analysis that doesn't list all four minerals is generally worthless for determining the true copper status of a ration.

Some additional points: The liver stores excess copper and then releases it for excretion or metabolic use. Blood is just the transport system. Insufficient dietary copper may indeed reduce blood copper levels, but ironically, high levels of dietary copper will *not* increase blood copper, at least at first, because the liver sequesters extra copper away from the blood until the last stages of toxicity. Although blood copper levels may be useful for detecting a copper deficiency, they are useless for predicting a copper toxicity.

Finally — and this is *the* well-known sheep fact — sheep are particularly sensitive to CCT because, over time, sheep tend to accumulate copper in their livers faster than other livestock species. CCT ends in metabolic disaster. When liver copper reaches a toxic threshold (usually above 800 ppm dry weight), the liver suffers catastrophic damage and dumps huge amounts of copper directly into the blood, causing a sudden rise of blood copper levels and the classic *hemolytic crisis*, which includes jaundice and death.

So sheep producers hear one recommendation over and over: feed "sheep salt" to their sheep rather than a standard trace mineral mixture designed for beef cattle. Actually, sheep salt is really just a standard TM mixture *without any added copper*. This recommendation is okay in areas where forages contain sufficient copper (or low levels of molybdenum), because TM mixtures designed for beef cattle all contain copper, and if these mixtures are fed to sheep, they may add too much copper to the sheep ration. But many producers, especially those who run both sheep and cattle, have historically never stocked two different minerals. For years, they've fed the same beef cattle TM mixture to *all* their livestock without incident. But things are changing in the cattle world, and this feeding strategy may now cause problems for sheep.

Among its many metabolic roles, copper is used by a number of enzymes in the immune system. Over the past ten years, researchers have observed better immune responses in dairy cattle that were fed diets containing more than 10 ppm copper. In response to this data, the new 2001 NRC (National Research Council) *Nutrient Requirements of Dairy Cattle* recommends 12–16 ppm copper in the total diet for dairy animals. In an analogous situation with beef cattle, some university bulletins now recommend that free-choice TM mixtures for beef cattle should contain *1,000–2,000 ppm copper*, and I've even seen recommendations as high as *5,000 ppm*. Some feed companies have already increased copper levels in their cattle mineral mixtures.

What do these new recommendations mean for people who raise sheep? Simple — it's now much riskier to offer sheep any mineral mixtures designed for cattle. Some cattle mineral mixtures may now contain much higher levels of copper than in previous years. If sheep consume these new cattle mineral mixtures containing copper at 1,000 ppm or more, well, quite a few sheep producers may become distressingly acquainted with the details of copper toxicity.

I say *may*, because not all cattle minerals contain these higher levels of copper. It depends on the company and the region. Usually, these higher copper levels are listed on the feed tag, maybe even with a warning "Do Not Feed to Sheep." But not always.

What about the neighbor who says, "Heck, I've fed the same mineral to my sheep and cattle for years, and I've never seen a problem"? Well, that was then and this is now. Back then he probably fed trace mineral mixtures with lower levels of copper. In today's world, commercial mineral mixtures may routinely contain very high levels of copper.

Today, any shepherd who wants to feed a beef cattle mineral to sheep should read the feed tag *very* carefully, and also know the background levels of copper, molybdenum, sulfur, and iron in the rest of the diet.

What about operations which run *both* sheep and cattle? Unlike sheep, cattle can handle the extra copper and are fairly resistant to CCT. But if we must feed both species at the same time, we do have some options. First, know the mineral levels in the total ration. Then you can decide which minerals are needed in the supplementary TM mixture and which ones should be avoided. Secondly, plan to feed two different mineral mixtures — one to the cattle, one to the sheep. This plan, of course, may be easier said than done, given the complexities of running a ranch or farm. Thirdly, look at other options for supplementing copper to cattle. Veterinary supply houses now carry slow-release boluses that contain tiny needles of copper oxide. These boluses lodge in the rumen and release their copper over six months or more. The boluses aren't cheap, but they may be less expensive than extra fence or labor, or losing sheep.

The alternative — of simply feeding a cattle TM mixture to sheep and taking a wait-and-see approach — is *not* a good idea. Remember that CCT develops slowly. Excess copper insidiously accumulates in the liver over months or years, with no overt symptoms. By the time the first sheep shows a hemolytic crisis, all the other sheep in the flock may also have livers loaded with copper and are just waiting for the other shoe to drop. Copper toxicity in sheep is very bad news. Avoid it at all costs.

First Published: October 2001

Author's Note: I've been in some areas that have avoided copper in their minerals for so long that their sheep are actually copper deficient. This situation presents a conundrum to the nutritionist. So I usually take the approach of formulating carefully (new mineral mixtures), measuring (liver samples), and monitoring (know the mineral consumption, lots of observations). Liver samples keep well in the freezer until they can be analyzed. But make sure they are labeled well.

Copper Extras

It's been awhile since we've discussed copper, especially chronic copper toxicity (CCT). I've written two previous articles about copper: one about its basic physiology and toxicity, and the other about specialized feeds such as sheep salt and beef cattle mineral mixtures. But there is more to this story. Copper nutrition is complex, and the risks of CCT depend on more than just feeding sheep salt. So let's tidy up some loose ends and examine other potential sources of extra copper.

If you've ever read agricultural discussions on the Internet, you might conclude that copper was the most horrible toxin known to sheep and that no one should ever feed it. Well, that's not quite true.

Copper is *required* by all mammals. Sheep need it at 7–11 ppm of their diet (from the 1985 NRC). But *excess* copper accumulates in the liver, and some genetic lines of sheep have trouble getting rid of it. Over time, liver copper levels can rise to a toxic threshold which severely damages liver cells. Those cells then break open and release their copper into the blood, causing a catastrophic *hemolytic crisis* which is usually fatal. Shepherds in toxicity-prone areas generally know about this problem and routinely feed their animals *sheep salt*, which is a trace mineral mixture that contains no added copper. So far, so good.

But this feeding strategy is not foolproof. Some sheep operations still experience CCT problems, which we all find unsettling. As I travel around the country giving workshops, someone usually comes up to me and describes a CCT problem on their farm. They want to know why. If they offer sheep salt to the animals, where is the extra copper coming from? Let's examine some possible sources . . .

The most obvious potential source is the mineral mixture. Sheep salt contains no added copper, and the small amount of copper that may find its way into the mixture from molasses or other ingredients is usually not a problem. The real problem is the use of *beef cattle mineral mixtures* — either because these products may be less expensive than sheep salt or because a producer is running both beef and sheep on the same property.

I know, I know. There are many old-timers who swear by their beef cattle minerals, saying that they fed these minerals to all their cattle and sheep for over 200 years and never saw a problem. That may be true, but times they are a-changing. The beef cattle mineral mixtures used years ago are not the same as many mineral mixtures formulated today. Research over the past 10–15 years has demonstrated that higher levels of copper for cattle can improve their ability to resist disease, and this makes sense since copper is involved in the immune system. Based on these studies, many Extension bulletins now recommend much higher levels of copper in cattle minerals. Many beef cattle minerals now contain 1,000 ppm copper or more. I've seen mixtures containing more than 4,000 ppm. I can assure you that the old-timers did not feed *these* minerals to their sheep, at least not for very long. I'd say that this is a compelling argument for carefully reading mineral feed tags, especially for beef cattle minerals.

And speaking of feed tags, did you know that very high levels of copper can improve growth in hogs and poultry? The commercial hog and broiler industries often include as much as 250 ppm copper in grower diets. We don't really understand the biochemical mechanism for this growth response, since copper is neither an antibiotic nor an estrogen, but from our practical shepherds' perspective, most of this copper passes through digestive tracts into the *manure*, and therein can lie a problem.

Let's say that a few miles away from your farm, a confinement hog operation raises a quadzillion young pigs. The managers routinely include 250 ppm copper in the grower rations. A quadzillion hogs produce a lot of hog manure. Quite a lot. And where do the managers put it? On every farm that accepts it — probably any place within 20 miles. Month after month, year after year. Copper in the manure becomes copper in the soil. What about the forages that grow from that soil?

You can't *automatically assume* that this manure will cause high forage copper levels, so you should conduct some tests. But don't test the soil for copper. There are too many intricate steps between field application and plant absorption to support a simple cause-and-effect assumption. If you are worried about manure causing high copper levels in the forage, test the forage that the animals actually eat.

By the way, does this mean that *all* hog and poultry manures are high in copper? No. Only manure from operations with *young growing animals*, and *only* if their feeds contain high levels of copper. Never from farms with layers or breeding sows. Another knot you must unravel.

But let's not forget that any high-copper feeds — for hogs or broilers or beef cattle — must originally be mixed in a *feed mill*. This is another potential source of excess copper. We trust our feed mills, but sometimes unexpected things happen. Employees may first mix a batch of hog feed and then mix your sheep feed next. Ideally, mills should flush out their equipment between batches, but the world doesn't always follow ideal procedures. Or perhaps someone makes a mistake and grabs the wrong premix bag. Or maybe someone mixes a feed batch for too long and the ingredients begin to separate, causing the heavier minerals to settle at the bottom of the equipment, ready to be picked up in the next batch. Or . . . well . . . you can imagine lots of scenarios.

These are the saddest situations because no feed mill wants this to happen, and innocent folks get caught in the middle. Sheep suddenly begin to die of CCT, some evidence points to a purchased feed, and people get angry. This is a time for patience and careful records. In fact, without good records on feed offerings, daily intakes, batch numbers, feed tags, timely veterinary involvement, and good feed and water samples, this situation is very hard to untangle. Especially since CCT takes a long time to develop (that's why this syndrome is called *chronic*), and copper began to accumulate in those livers months or years earlier. Indeed, the actual source of copper may not have been the commercial feed at all. I urge anyone who suspects CCT to keep good records and involve a veterinarian immediately.

Finally, there is *fruit*. As in pears, apples, and grapes. As in orchards and vineyards. Why? Because for over a century, one of the most reliable sprays against leaf fungus has been *Bordeaux Mixture*, which is composed of copper sulphate, hydrated lime (calcium hydroxide), and water. Some orchardists use alternatives called *Fixed Copper Sprays*. In either case, the effective chemical agent is copper. Over time, this copper moves from the trees into the soil and possibly into the forages.

It's easy, of course, to pinpoint this situation if your sheep graze in your own orchard. You know if you've sprayed your own trees with a copper compound. But things get a bit fuzzier on rented fields, where sheep can move under trees that someone else may or may not have sprayed. And then there are pastures that *used to be orchards*, years and years ago, that have been converted to grass. And these pastures may have been rented from a current owner who doesn't even know about those old orchards . . .

It comes back to testing and good records. If you have any doubts about a pasture, simply test its forage for copper (and molybdenum and sulfur and iron, which can all influence copper availability). Maybe run tests at different times throughout the year, just to be sure. And take care with your sampling procedure — you need to be confident that the laboratory results from your sample reflect the true conditions of an *entire* field.

Of course, we may encounter other sources of extra copper — oddball one-of-a-kind situations where we scratch our heads in amazement — but the sources I've just described are the major ones. Copper nutrition is indeed complex, and we should always watch for the extra sources that can make things even more complex. In the end, copper does not have to be the devil in sheep's clothing.

First Published: October 2006

Author's Note: Surprisingly, the 2007 Small Ruminant NRC tends to recommend lower copper levels in sheep rations than the 7–11 ppm recommendation of the 1985 Sheep NRC. Which recommendation works better in the realities of farm and ranch operations? Which potentiates problems of toxicity or deficiency? An interesting speculation — time will tell.

Iodine (1)

Something weird happened in western Oregon this year. For the first time in memory, our livestock experienced widespread problems with iodine deficiency. Iodine? Yeah, iodine. Unmistakable symptoms: lambs and kids born weak or dead, with goiters, often without hair or wool. Wait a moment. How could this happen? Doesn't everyone include iodine in their trace mineral mixtures, and don't most commercial mixtures contain at least 70 ppm iodine? Well, yes, but nonetheless we're still seeing these problems. Hmmm . . . maybe iodine nutrition is not as simple as we think.

First of all, mammals (livestock, us, etc.) use iodine for one main purpose — to form the iodine-containing compounds associated with the thyroid gland — a tiny organ in the throat region. Most people know two things about iodine: it is somehow involved with the thyroid hormone *thyroxine*, and its deficiency causes *goiter*, which is an enlarged thyroid gland.

Humans have been coping with goiter for a very long time, even if they didn't always use chemical jargon to describe it. The Chinese knew that certain marine items were beneficial to the thyroid gland, and the ancient Greeks relieved goiter by adding seaweed and burnt sponges to their diets. A crude delivery system, perhaps, but it worked. After scientists first identified iodine as an element in 1811, they soon found that those traditional cures contained iodine. The first folks to prevent goiter by adding iodine to salt were the Austrians in 1895. The Americans followed suit in 1924, and by the 1960s most countries in the world were routinely providing iodized salt to their populations.

Some physiology: The thyroid gland contains more than 70% of the iodine in the body. When iodine is absorbed from the gut, the blood transports it to the thyroid gland where enzymes combine two atoms of iodine with the amino acid tyrosine to form *diiodotyrosine* (also known as *T2*). Two of these molecules are then fused to form *tetraiodothyronine* (*T4*). This is the molecule we call *thyroxine*. Thyroxine then moves from the thyroid gland into the blood, but it's not quite ready for prime-time. As it passes through other organs, thyroxine is activated into a more potent molecule by enzymes that remove one of its iodine atoms to form *triiodothyronine*, which is also known, logically, as *T3*.

The T3 molecule is the actual hormone that exerts powerful control over some of the most fundamental metabolic functions of the body. It regulates the basal metabolic rate, which means the underlying rates of heat production and protein turnover. It's also involved with muscle function, fetal development, thermal regulation, reproduction, growth, and the immune system. Whew!

Thyroid compounds are so important that the body has developed sophisticated feedback mechanisms to cope with deficiencies. If the brain detects low blood levels of thyroxine, the pituitary gland secretes additional TSH (*Thyroid-Stimulating Hormone*) which tells the thyroid to get cracking and produce more thyroxine. But if iodine is in short supply, the thyroid tries to compensate by expanding cell size and making more cells, which increases the size of the gland. This enlargement is what we call *goiter*. Sometimes a goiter can be huge and protruding — lots of textbook photographs attest to this — but a goiter can also be small and only detectable if you palpate it with your fingers. Since the normal thyroid gland is quite tiny, you can assume that if you feel anything hard with your fingers, it's probably a goiter. And by tiny, I mean tiny. A normal thyroid gland in an adult cow weighs only four grams — only slightly heavier than a dime.

But if we depend only on goiters to alert us about deficiency problems, we may be overlooking other symptoms. In the growing fetus, iodine is involved with development of the brain and also skin, hair, and wool. When a pregnant female is deficient, she goes into an iodine conservation mode, which includes reducing the amount of iodine transferred across the placenta to the fetus. She effectively saves herself at the expense of her fetus. Thus, our first hint of deficiency will usually occur in newborns — symptoms such as animals born weak or dead, often with goiters (which may not be large), with thin or pulpy or unusual skin, with a reduced haircoat or even no wool or hair at all. Or "dumb" animals that refuse to suckle.

Deficiency symptoms are not restricted to newborns. Older animals can suffer from reproductive problems — infertility, irregular estrous, low libido, etc. — or slow growth, poor adjustment to cold weather, or increased mortality. Do any of these symptoms sound familiar?

Iodine deficiency, however, may not just be caused by a straightforward lack of iodine in the feed. The world of livestock nutrition is not a simple place. Some forages contain compounds called *goitrogens* that interfere with iodine metabolism. Goitrogens prevent the thyroid from properly absorbing iodine from the blood or incorporating it into thyroxine. The brassicas (rape, kale, turnips) can contain high levels of *glucosinolates* which are compounds that give these plants a "hot" taste and may help them to cope with infectious fungi, but these compounds are also goitrogens. One of the standard caveats about feeding brassica forages is to make sure that the animals have access to lots of iodine.

Rape (*Brassica napus*) is particularly important because rapeseed is often a profitable crop. Unfortunately, rape can contain glucosinolates and also another toxin called *erucic acid* (which causes heart lesions in non-ruminants). But all is not lost. Canadian researchers developed a type of rape bred specifically for low levels of both compounds. They derived its name from the phrase "Canadian Oil Low Acid" — *Canola*.

An extra twist to the goitrogen story is that any feedstuffs which contain low levels of cyanide compounds — like sorghum-sudangrass, white clover, or raw soybeans — may cause iodine problems. Why? Because animals detoxify cyanide by converting it to a compound called *thiocyanate*, which is a goitrogen.

Remember the conversion of thyroxine to T3? Well, the enzyme that converts T4 to T3 is a selenoprotein, which means that it contains *selenium*. Which means that low dietary levels of selenium can potentially reduce the formation of T3. For example, a high rate of sulfur fertilization can reduce forage selenium levels and cause a mild selenium deficiency — perhaps not enough to result in the classic symptoms of white muscle disease, but enough to reduce the T4 to T3 conversion rate and cause a secondary iodine deficiency. In fact, in Europe, iodine supplementation is often sold together with selenium.

Interestingly, even though iodine is important to animals, our forage plants don't actually need it. Plants contain no metabolic systems that include iodine, so any iodine that shows up in their tissues has simply crossed into the roots along with other soluble compounds. Two critical factors that influence plant levels are geography and geology.

Since iodine is abundant in seawater, it follows that rainfall originating over an ocean will contain quite a bit of iodine. People have historically known, and research has substantiated, that soils near the ocean usually have sufficient iodine, while interior soils are usually iodine-deficient. But geology also plays a role, because even interior soils can be high in iodine if they developed from ocean sediments, unless of course those soils were scraped away by glaciers.

Things do get complicated. But in general, plant levels of iodine usually occur in the same range as selenium (another element that plants don't need) — between 0.04 ppm and 0.4 ppm of the dry matter.

Now we have some background. Next month, we'll look at the details of what happened this year in western Oregon, and why, and what we can possibly do about it.

First Published: June 2005

Author's Note: "This year" refers to the winter of 2004–2005. In western Oregon (the "rainy" section west of the Cascade Mountains), forages begin their growing season with the onset of the fall rains in October or November. Western Oregon experiences a mild, maritime climate with mild winters of few frosts, and then an early spring, with its legendary rains throughout this period. Grass growth begins rapidly in the fall, slows down through the winter and then explodes in the spring, finally going reproductive and dormant soon after the rains cease in May or June. Legumes grow in the fall, go into dormancy during the winter because the soil temperatures drop below their minimums, but then show explosive growth in the spring. The summers are warm and bone dry, with no forage growth possible without irrigation.

Iodine (2)

How much iodine do animals need, and why can't they sometimes get enough of it? And specifically, why during the past year did many experienced sheep and goat ranchers in western Oregon observe severe iodine deficiencies in their animals? Iodine nutrition is complex, so let's continue our discussion from last month.

Let's address the first question first: iodine requirements. Livestock need approximately 0.1–0.8 ppm iodine in the total diet. The various NRC publications list slight variations between different species, but at least for sheep, the NRC specifically mentions that the upper end of this range applies to gestation and lactation. Iodine toxicity is usually not a problem. Most livestock can tolerate relatively high amounts of iodine, up to 50 ppm. But not horses. Horses are particularly sensitive to iodine toxicity, and the NRC recommends that their diets contain no more than 5 ppm.

Iodine is an integral part of thyroxine, a hormone which controls basal metabolic rate and heat production. Iodine requirements vary, however, in response to temperature. As you'd expect, cold weather increases the need for thyroxine because animals must produce more body heat. I know, I know — readers in Minnesota or Saskatchewan may snicker at our western Oregon definition of "cold weather," but ranchers here typically lamb on pasture during midwinter, often in a driving rain at 35°F. Our climate may not be cold enough for ice-hockey, but it's definitely cold enough to elicit a thyroxine response. And one symptom of iodine deficiency is that newborn animals have trouble surviving in cold temperatures.

Last month, I described how goitrogenic compounds can interfere with iodine metabolism — things like *glucosinolates* in brassicas and low levels of *cyanide-containing compounds* in sorghum-sudangrass and white clover. These goitrogens can increase iodine requirements from two- to tenfold, which means that animals grazing brassicas could need 2 ppm iodine in their total diets rather than 0.2 ppm.

The good news about iodine metabolism is that animals are very efficient at absorbing iodine from feeds. Animals also conserve iodine by recycling it — they secrete unused iodine into the *abomasum* (true stomach) and then re-absorb it downstream from the small intestine.

Forage levels of iodine are all over the map. Literally. Some soils are naturally high in iodine; some are not. Iodine is abundant in seawater, and soils derived from marine deposits can contain quite a lot of iodine, although glaciation can remove iodine-rich topsoils. Any rainfall that develops over an ocean contains iodine which will be deposited in the topsoil. Forages grown near the ocean can be high in iodine, but levels decline with increasing distance from the sea. Also, mature forages contain less iodine than younger plants, and grasses tend to be lower in iodine than legumes grown on the same soil.

In contrast, any feedstuffs derived from the ocean are high in iodine. Seaweed, for example, can contain *more than 4,000 ppm* iodine. Livestock have no goiter problems when fed diets composed of crabmeat, fish meal, and seaweed, but since these are not exactly a common feedstuffs, we usually need to supplement iodine.

The North American strategy is to add iodine to trace mineral mixtures, generally at levels of 70–100 ppm, typically by using compounds like calcium iodate or potassium iodide or the organic compound EDDI (*ethylenediamine dihydroiodide*). All are highly absorbable by animals, although the soluble forms can sometimes be lost from minerals through leaching and volatilization if exposed to bad weather for extended periods of time. The rest of the world, however, often uses other methods to supplement iodine. In England, farmers can use slow-release iodine boluses. New Zealand farmers can inject their livestock with long-acting iodine in the form of iodized oil. And because human iodine deficiency is fairly common in developing countries, international aid organizations routinely give people iodized oil either by injection or by oral capsules. Unfortunately, none of these alternative delivery techniques are currently available in the U.S.

So . . . back to Oregon and our second question: what happened last winter and, most importantly, why? Well, a number of veteran producers with large herds and flocks observed lots of goiters in newborn animals. Once ranchers learned how to palpate the throat region with their fingers, they found many smaller goiters, which suggested that iodine deficiencies were more widespread than originally thought. Lambs and kids were born dead or weak, sometimes with pink skins without hair or wool. Often they were unwilling to suckle. Flocks experienced high mortality rates, especially in the colder weather, and in one goat herd, kidding was extended out over a much longer period than normal.

Looking for causes, we can list the usual suspects: (1) simple lack of iodized mineral mix, (2) quality-control problems in the minerals, (3) grazing brassicas or other plants containing glucosinolates, (4) grazing other forages with a cyanide potential such as white clover and sorghum-sudangrass, and (5) subclinical selenium deficiencies which interfered with the conversion of thyroxine to T3.

These reasons all sound plausible and scholarly — but none of them stand up well against some critical facts. The most significant fact is that these ranchers all have different management systems. Although they all grazed their animals during the winter, their fields were quite different from ranch to ranch, and few animals grazed fields containing brassicas or other goitrogenic plants.

In addition, all these ranchers were very experienced — none would let their minerals periodically run out. No one fed simple white salt, and all their mineral mixtures contained iodine. Since their mineral mixtures came from different companies, it's extremely unlikely that *all* these mixtures had the same quality-control problem with iodine. We also eliminated selenium interactions. Every Oregon rancher knows about selenium supplementation, and most commercial mineral mixtures contained at least 85 ppm selenium, which is usually sufficient. Some of these operators also routinely gave selenium injections. In short, there seemed to be no simple common management denominator among these ranches.

But one nagging question kept coming up in my thoughts: why *this* year? Iodine problems were so rare in previous years that they were hardly noticed, and none of these ranchers had changed their management systems from previous years. So why did they all suddenly run into significant iodine problems *in the same year*? Hmmm. What factor could have been common to all these ranches and also unique to this year?

The weather, of course. And maybe, just maybe, that was the answer.

Everyone here is talking about this year's weather. Forage yields in western Oregon are higher than anyone has ever seen. Last fall was a "perfect" growing season, with early rains that were properly spaced, giving lots of moisture to the warm ground. Our winter was exceptionally mild, and there were plenty of rains during the late spring. The result — a huge growth of forage.

So how does this relate to an iodine problem? Because forage growth affects grazing behavior.

Remember that plants don't need iodine, so the small amounts of iodine found in forages are incidental to the vagaries of soil levels and root absorption. Although grazing livestock obtain some iodine from their forages, they actually can obtain more from the soil. Because rainfall enriches the topsoil with iodine, the topsoil contains more iodine than the plants that grow on it. Grazing animals tend to ingest soil, either directly or from particles splashed onto the forage. In most years, animals graze so close to the ground that they can't help but consume some topsoil.

But this year was different. Our phenomenal forage growth gave animals the luxury of *grazing the highly-palatable tops of the plants rather than the lower parts near the ground.* Which means that animals probably did not consume particles of topsoil and thus did not receive that extra iodine. And this situation occurred in nearly every ranch throughout the area. This grazing behavior, combined with the normal variation in mineral intake and the cold temperatures during pregnancy, makes for a compelling nutritional argument.

Interestingly, if you search through the scientific literature, you will find singular references to the observation that goiter problems seem to occur more frequently during "good years" rather than during years with average or poor forage growth.

So if the good weather was the problem, what can we do about it in the future? Well, we can hope for bad weather and poor forage years to force our livestock to eat topsoil — but somehow I don't think that's the best idea. Or we can go through our animals each week during pregnancy and paint iodine on their heads (iodine is absorbed directly through the skin). No, I don't think that's a practical idea either. Or we can carefully assess the growing season and, if forage conditions look very good, we can try to supply higher levels of iodine to our livestock and also get ourselves prepared for possible iodine problems at lambing or kidding or calving. This idea makes sense.

One thing is for certain: we will never again take iodine nutrition for granted.

First Published: July 2005

Author's Note: References to the "past year" and "this year" refer to the winter of 2004–2005. And as of this writing, there is a slight interest among official channels to obtain regulatory approval for the use of alternative methods of supplementing iodine to our livestock.

Selenium Pools

The selenium issue is a little like professional baseball — every year brings new developments, it's a source of avid discussion, and people rely on statistics to support whatever opinion they currently espouse. When discussing selenium, instead of quoting batting averages, folks usually quote selenium levels in the blood. Either way — baseball or selenium — numbers count. Numbers also confuse. Blood levels of selenium can be tested in different ways, and each number has a different meaning. So let's look at these numbers closely.

First, let's clear-up some confusion about blood terminology. We can think of blood as a kind of a soup composed of two parts: the liquid and the things floating around in the liquid. The things floating in the liquid are mostly red blood cells. The liquid is either plasma or serum, and the difference between the two is simple. After blood is collected, it is spun down in a centrifuge. The heavy cells settle at the bottom of the tube; the liquid rises to the top. If the blood is allowed to clot prior to centrifugation, the liquid is called *serum*. If the blood is prevented from clotting, the liquid is called *plasma*. The difference is that the plasma contains all the clotting factors and related components, while the serum does not. Finally, the term *whole blood* is, well, everything — all the red blood cells, clotting factors, and liquid.

In scientific publications, scientists tend to shy away from using the term "soup." Instead, they refer to the two parts of blood as two *compartments*, or two *metabolic fractions*, or better, two *metabolic pools*. Thinking about blood as two metabolic pools is a good mental strategy, because these two pools are chemically quite different, and measurements of each will give different results, as we shall see for selenium.

Between 67–80% of all blood selenium occurs *inside* the red blood cells, primarily as part of an enzyme called *glutathione peroxidase* (GSH-px). This is the molecule that helped shape our current feed laws. Until 1973, we only had a general idea that selenium was an essential nutrient, but we hadn't pinpointed a precise metabolic use for it, and federal regulations forbade its inclusion in livestock feeds. In 1973 researchers first reported that GSH-px was a selenium-containing protein, which proved beyond a shadow of a doubt that selenium was required in a metabolic pathway in livestock. Identifying an enzyme that contained selenium convinced the FDA to change its regulations to allow selenium in feeds.

Because GSH-px contains most of the selenium in the blood, researchers initially tried to use its enzymatic activity as an indicator of selenium status. But after analyzing thousands of samples, they found that values for GSH-px activity were notoriously inconsistent, especially between labs. GSH-px is an enzyme, and enzymes are delicate molecules. Enzyme activity depends on more than just selenium composition — factors such as temperature, reagent quality, handling procedures, technician skill, and also if the samples were left on the dashboard of a pickup for three days before being tested. There is no standardized assay for GSH-px. Values from one lab cannot be used confidently by another lab. This is not an ideal assay for diagnosing selenium status. True, GSH-px values can be very useful for experiments *within* a single lab over a short period of time, but their routine use for selenium diagnosis has fallen by the wayside in favor of other measurements.

Today, we typically rely on selenium values from either whole blood or from serum or plasma. (From a selenium perspective, there is little difference between serum and plasma, so for brevity I'll just refer to serum.) But these values also have very different interpretations.

Professionals don't agree on the precise selenium levels, but in general, adequate selenium levels in sheep run between 0.120–0.500 ppm for whole blood and 0.080–0.400 ppm for serum. That is, if you are in the United States. The *rest* of the world uses the metric system, so *their* selenium values often look different. One ppm equals one milligram per liter (mg/l), which is the same as one microgram per milliliter (Fg/ml). But because selenium values are so small, with so many zeros, some laboratories prefer to report selenium values as nanograms per milliliter (ng/ml) which is the same as parts per *billion* (ppb). Which is equivalent to taking ppm and moving the decimal three places to the right. Follow me? Anyway, I like these values more because they are larger numbers. Namely, 120–500 ng/ml for whole blood and 80–400 ng/ml for serum.

Now for the difference between the two pools of selenium.

Red blood cells, which contain most of the blood's selenium, have definite life spans. After 3–4 months, old red blood cells are destroyed and replaced by new cells. Selenium remains in a cell's GSH-px for the entire life of the cell, and this amount is fixed when the cell is formed. Therefore, the selenium value for *all* the red blood cells in the body is really a rolling average — a reflection of the selenium status averaged over different periods of red blood cell formation, including values for old cells as well as for new cells. This lag period shows up as a delayed response to supplementation or a delayed response to reduced selenium intake.

This lag period therefore applies to the selenium values of whole blood, because whole blood levels are overwhelmingly influenced by the amount of selenium in the red blood cells. For example, supplementing selenium to a deficient animal takes three months to maximize the selenium levels in whole blood.

Thus, the selenium level of whole blood depicts the relatively long-term selenium status of an animal. While this number may be useful historically, it may be quite misleading as an indicator of current events.

In practice, we are usually interested in the *current* selenium status of our animals because we need to make timely decisions about supplementing selenium or treating deficiencies. Serum selenium is a better gauge of this than whole blood selenium. Because serum is primarily a liquid transport system, not a final, long-term destination, serum selenium levels rise and fall relatively quickly in response to supplementation or shortage.

For example, supplementing selenium to deficient animals causes serum levels to rise quickly and reach maximum levels within weeks, while whole blood levels may show little change during this period. The slow response time for whole blood could easily mislead producers into thinking that their selenium supplementation was ineffective.

One interesting characteristic of serum selenium is that these levels respond slightly differently to different forms of supplementation. Inorganic forms of selenium (selenates or selenites) cause serum levels to plateau at lower values than the organic forms (seleno-amino acids). Both types of supplementation can be effective against selenium deficiencies, but remembering how serum responds may be useful when looking at marketing claims or trying to interpret laboratory reports.

So which selenium value to use? Which to believe? Actually, the *best* place to test for selenium is the liver, not the blood. But in a practical sense, liver is a rather inconvenient tissue to sample, so we are stuck, so to speak, with blood. No problem, as long as we understand which value is which, and we don't try to "pool" our samples.

First Published: December 2002

Author's Note: In the field, I often hear veterinarians, producers, and nutritionists speak in kind of a heated babel about blood selenium levels, but when I ask about the details, about which assays were done, almost no one knows. (In contrast, baseball aficionados can often quote chapter and verse about *their* statistics.) But recent developments in selenium nutrition — adding selenium to dry fertilizers, the increased availability of organic forms of selenium in feed supplements, and the growing number of commercial forage laboratories offering affordable selenium assays — make it even more compelling for us to know which numbers we are talking about and what they mean.

Selenized

The selenium world has suddenly become more complex. This spring the Oregon Department of Agriculture began allowing fertilizer companies to add selenium to dry fertilizers to increase the levels of forage selenium. For livestock operators in our selenium-deficient region of western Oregon, this would seem like a wonderful breakthrough. Or is it? We now must deal with complexities that we never dreamed of — entanglements that everyone may eventually face.

First let's review some numbers. Livestock require selenium at approximately 0.1 ppm in their total diet, with 0.3 ppm being better. Selenium levels over 3 ppm can cause toxicity. This is a rather narrow window between requirements and toxicity. Ironically, forages do not require selenium. Forages grow perfectly well without any selenium because they have no metabolic systems that use it. But since selenium chemically resembles sulfur (selenium is directly beneath sulfur in the Periodic Table — we all remember *that*, of course), plant roots will absorb selenium from the soil, and plant cells will incorporate it instead of sulfur into amino acids. Therefore, forage selenium levels reflect the amounts available in the soil, and in selenium-deficient regions, that's not very much. Forages grown in deficient areas typically contain less than 0.1 ppm, often even lower than 0.05 ppm.

For decades, livestock owners in the U.S. have routinely coped with this situation by feeding supplemental selenium, either in trace mineral mixtures or in supplements. The FDA sets the maximum selenium levels in feeds, and they are currently 90 ppm (for sheep) or 120 ppm (for beef cattle) in trace mineral mixtures, or 0.3 ppm in the total diet for both species. One interesting loophole in the feed laws, however, allows veterinarians and feed manufacturers to add higher selenium levels into mineral mixtures as long as those mixtures are labeled "*premix*" with feeding instructions that call for adding these premixes to other components, like salt. I've seen selenium levels in premixes of over 200 ppm.

These strategies work well, as long as animals consume these feeds properly. With good selenium supplementation, farmers don't need to use selenium injections. So far, so good.

But in New Zealand, farmers have taken a *different* approach to selenium supplementation: they *don't* put selenium into salt mixtures. Instead, for the past twenty years, they have been allowed to *add it to their fertilizers.* This strategy works for them — because the resulting forages contain enough selenium to meet livestock requirements.

Now back to Oregon. Over the past few years, Oregon ranchers and university staff have cooperatively experimented with a similar fertilizer strategy, carefully monitoring selenium levels in the forage and in the blood of animals grazing that forage. They've demonstrated its effectiveness to the Oregon Department of Agriculture, which this spring agreed to permit it. For the first time, Oregon farmers and ranchers can now apply fertilizers that provide 4.5 g selenium per acre. And some are already doing it.

The selenium product used by fertilizer companies contains a combination of *sodium* selenate and *barium* selenate. Barium selenate is less soluble than sodium selenate, which means that its selenium becomes available to plants over a longer period of time. In effect, this commercial product combines a fast-release selenium and a slow-release selenium, and it boosts the forage selenium levels for a year. How high? Lots of variability here, but selenium levels generally rise to 0.2–0.5 ppm, and I've seen reports of 0.6 ppm. These levels are sufficient to meet the needs of livestock, and they are well below toxic levels.

As an added benefit, selenium atoms in these forages are in amino acids rather than in selenates or selenites common to mineral mixtures. Amino acid selenium is more available to animals than the inorganic mineral forms.

Now the complexities.

The first and most obvious: what is the risk of toxicity if a rancher uses the new selenium-enriched fertilizer *and also* offers a mineral mix containing selenium? The typical recommendation will be to discontinue such offerings, but there may be times when extra selenium is fed anyway. Some producers may not want to change their mineral product so quickly, or they may not trust the fertilizer efficacy. Then what?

To answer this, we need to know a few details: the selenium level in the forage, the selenium level in the mineral, the daily intakes of each, and most importantly, the *toxic dosage* of selenium to livestock. Recognize that a toxic dose is not just a standard amount. For example, 2 milligrams of selenium in a baby lamb is quite different physiologically than 2 milligrams in a large cow. In the world of toxicology, toxicity is properly expressed as an *amount per body weight.* For selenium, an extended dose for long-term toxicity is 0.08 mg per kg body weight.

So let's run the numbers. We can do the calculations by hand, of course, but why do that when there are spreadsheets? So I created a simple spreadsheet, put in the formulae, and ran dozens of "what if" scenarios. I varied forage selenium levels from 0.2–0.7 ppm, forage intakes from 2–4% of body weight, and mineral supplements containing 90, 120, and 200 ppm selenium at various intakes. Here is what I found:

In general, sheep consuming forages that contain up to 0.7 ppm selenium, at even 4% of body weight, probably won't run into toxicity problems if the supplemental mineral contains 90 ppm or less. Unless, of course, those animals consume the TM mixture at higher levels than expected. (For example, 2 ounces per day instead of 1 ounce.) Cattle are probably okay in all the equivalent scenarios because they don't eat as much forage as sheep, as a percentage of their body weight. Cattle will never consume over 3% of their body weight in forage, while sheep will often eat more than 4% of their body weight in very high quality grass.

But I found that sheep *can* run into problems if they consume large amounts of 90 ppm mixtures, even if the forage contains only 0.3 ppm selenium. This is disturbing because some TM mixtures contain ingredients like molasses that encourage consumption, and their real-life intakes can indeed be greater than 1 ounce per day. Therefore, with sheep, it may be prudent to discontinue the use of a 90 ppm high-intake mineral if selenium-enhanced forages are fed.

However, *a real toxicity risk occurs with those special high-selenium premixes* — mixtures containing 200 ppm selenium. Even with a forage level of 0.3 ppm selenium (a typical and moderate increase from the fertilizer), and a modest mineral consumption of only 1 ounce daily for sheep, the selenium dosage approaches toxic levels.

The bottom line here is that any producer who applies selenium-enriched fertilizer should never feed sheep a high-selenium *premix* when feeding that forage. For lower mineral levels (90 or 120 ppm), the toxicity risk is probably significant only if the mineral intake is very high, *combined* with high intakes of selenium-enriched forage. Cattle are less susceptible because they consume forage at lower rates. But for either species, producers should *not* use selenium injections if they feed selenium-enhanced forages.

Another thing to consider — what about *purchased hay* — on either side of the transaction? If I *buy* hay, I would now want to know if the grower had used selenium fertilizer on that land, and when, or else be prepared to have that hay analyzed for selenium. Otherwise, how could I know if it was safe to feed selenium-containing minerals? The good thing here is that animals consume more high-quality pasture than hay, so the contribution of hay to the selenium dosage would not be as high. But the flip-side of this problem is that if I *sell* hay, should I now include a user's guide with my hay about its selenium fertilizer history? Am I contractually liable for possible toxicity problems on someone else's farm?

Finally, I need to mention that most reference values for selenium requirements and toxicities assume that animals consume selenates or selenites, not the more available seleno-amino acids. Forage selenium is primarily in amino acid form, and if forages constitute a major proportion of the diet, then those reference values may be too high. Which means that the long-term toxicity dosage may actually be *lower* than the reference value of 0.08 mg/kg body weight, or not.

Well, I guess there are few wrinkles to work out yet . . .

First Published: June 2003

Author's Note: The risks of long-term selenium toxicity is a complex issue. As of this writing, there are some straightforward research projects that still need to be conducted — hypotheses involving real-world duplications of potentially toxic situations, especially with sheep and goats who generally have a higher forage intake than cattle. Although we are seeing routine field recommendations to use selenium in fertilizer — because it generally works — it's a little like starting down a steep hill without really knowing if your brakes are okay. To my mind, that's a bit unnerving.

Kesterson

In the 1980's the Bureau of Reclamation and the U.S. Fish & Wildlife Service agreed to cooperate on a wetlands project in California. And because of that agreement, in 1993 the FDA tried to reduce the selenium levels in animal feeds by more than 60%. This is the story of how those two events are linked.

The western area of California's San Joaquin Valley receives an annual rainfall of less than 10 inches — similar to Alice Springs in Australia, and less than Jerusalem. Growing crops means adding water. To help farmers, the federal Bureau of Reclamation did just that: it built an extensive series of canals to deliver fresh water to the San Joaquin farmland in Fresno County. So far, so good.

However, irrigation is not a simple matter of dumping water on the land, stepping back, and watching crops grow. Farmers irrigate to survive, but in dry climates farmers also fight a constant battle to prevent salt intrusion into the root zone. Beneath those Fresno County farms is a pan layer of impermeable soil that prevents water from draining deeper into the ground. If farmers pour irrigation water onto the soil surface without sufficient drainage, the ground becomes increasingly waterlogged. Water in this soil allows soluble minerals to migrate towards the surface. Typical of dry climates, those San Joaquin soils are alkaline (pH > 7.0), and the subsoil layers contain relatively high levels of various minerals, including selenium.

The Bureau of Reclamation devised a clever solution: it would deliver fresh irrigation water to the soil surface and also build a series of subsurface irrigation pipes. The Bureau would then flush the excess water and minerals into drainage ditches and drain the entire runoff into the San Joaquin River delta which ultimately flows into the ocean. Along the way, it would also build a few shallow evaporation ponds to control seasonal water flow.

And so, in the early 1970's, the Bureau buried perforated subsurface drainage pipes beneath that fertile Fresno County farmland. These pipes were larger versions of the common tile pipes that private farmers used to drain wet fields. The Bureau installed large sump pumps to pull the water from the soil into those pipes, and it built the San Luis Drain to carry this subsurface water away from the farms. Downstream next to this drain, the Bureau also dredged out a series of twelve shallow ponds to serve as evaporation pans and holding basins for this water.

These twelve ponds were collectively called the *Kesterson Reservoir*. The ponds averaged only 3–5 feet deep with a total surface area of about 1,100 acres, and with their surrounding grasses and sedges they formed a seasonal wetland. Fish and insects quickly populated the shallow waters, and birds were attracted to these wetlands for feeding and nesting. Although this project was fairly expensive, the wetlands habitat gave the Bureau of Reclamation an attractive option: to offset some costs, the Bureau agreed to donate these wetlands to the Fish & Wildlife Service. The entire area of the twelve ponds and their surrounding marshes then became officially known as the *Kesterson National Wildlife Refuge*.

But all did not go well. By 1983, stories began to surface of biological problems: dead fish, low egg hatches, deformed embryos and hatchlings, dead birds. The evidence pointed to selenium toxicity.

Why selenium? Let's look more carefully at those San Joaquin soils. Inorganic selenium comes in four forms: selenate, selenite, elemental selenium, and selenide. Selenate is very soluble and flows with the water. Selenite tends to adsorb (yes, that's a "d" not a "b." It means to stick on the surface) onto soil particles but it can be displaced into the surrounding water under certain conditions. The other two forms of selenium are insoluble and rather immobile. The problem was that the dry subsurface soils beneath that California farmland contained relatively high levels of selenium, which in that alkaline environment were mostly in the *mobile* forms of selenate and selenite. By adding large volumes of irrigation water to the soil surface and pumping the subsurface water into drainage ditches, the Bureau of Reclamation had in effect created a highly effective rinse cycle that swept subsurface minerals into the drainage water. Including selenium.

Selenium concentration is usually expressed as parts per *million* (ppm) or even parts per *billion* (ppb). These are very low levels, but as we all know, sheep and cattle require selenium only in very small amounts. As little as 0.1 ppm selenium in their diets will prevent deficiency symptoms. The toxicity window, however, is rather narrow. Toxic levels in livestock diets begin at around 2 ppm.

In nature, selenium levels in water are usually quite low. Typical surface water usually contains less than 2 ppb selenium. The Kesterson waters, however, contained much higher levels of selenium: water in the San Luis Drain contained *300 ppb selenium*, and water in the lower Kesterson ponds contained over *50 ppb*. Fish and birds from the Kesterson wetlands showed elevated levels of selenium in their tissues. Something horrible was happening. Selenium that normally would have remained safely in the ground was being systematically pumped into the drainage ditches.

Reporters learned of the Kesterson situation and publicized it with stark photographs and alarming books. Public environmental groups complained to the government. A few private testing companies lobbied for more frequent selenium testing of animal feeds. And in 1992, the FDA held formal hearings to reassess its 1987 ruling which permitted higher selenium levels in animal feeds. The FDA feared that the higher feed levels could be redistributed through manure and urine, and thus ultimately raise selenium levels in the environment, particularly in surface waters.

In September 1993, the FDA published its reconsidered ruling: reduce selenium levels in animal feeds back to pre-1987 levels (with a year's grace period to readjust manufacturing processes). This meant a 67% reduction of selenium levels in complete feeds and trace mineral salts for sheep and an 83% reduction in trace mineral salts for cattle.

The main evidence used to support this ruling was the story of the Kesterson Reservoir — the story of a government agency that had systematically pumped subsurface drainage water from an alkaline, high-selenium soil into a stagnant wetland. Virtually no evidence was presented to the FDA that dietary selenium in animal feeds significantly raised selenium levels in the environment.

So how, you may logically ask, did the FDA equate the Kesterson evidence with the realities of feeding animals in the wetter regions of the U.S. where the soil pH is usually acidic and selenium is deficient?

Beats me.

First Published: August 1994

Author's Note: Because of this brouhaha, the FDA never implemented those 1993 roll-back regulations. During the public input period, scientists, feed manufacturers, and farmers and ranchers raised enough of an outcry to attract considerable attention. They wanted the FDA to follow scientific evidence, not politically-charged headlines. Congress heard their complaints and agreed. In 1994 Congress expressly forbade the FDA from overturning the 1987 regulations until it had enough scientific evidence to justify reducing the feed levels of selenium in feeds and mineral mixtures. As of this writing, the 1987 regulations are still in effect. And what happened to the Kesterson Reservoir? Well, in the late 1980s, the ponds were closed and drained.

FDA's Time Machine

Have you read the papers lately? Or checked with your feed store? The FDA, in all its infinite wisdom, has accomplished a feat that scientists and fantasy writers have long wished for — it crossed the time barrier and turned back the clock. Not quite to the Jurassic Period, but almost as grim. Welcome to 1986.

In case you haven't heard: on September 13, 1993, the FDA published a new ruling about selenium. In one blinding stroke of the pen (actually, in 12 depressingly long pages of the Federal Register), the FDA reduced the permissible levels of selenium that could be included in feeds, feed additives, and mineral mixtures. To be precise, the FDA "stayed" the 1987 revisions of the original regulations, which means, in regulatory gobbledygook, that the 1987 laws are temporarily put on hold, kind of like in a penalty box, and the pre-1987 regulations have come back into effect. As I said — welcome to 1986.

This new ruling affects everyone who raises livestock in selenium-deficient areas — like all my neighbors in western Oregon and everyone else who has ever experienced white muscle disease on their property.

A little history: In 1987, in response to a petition from the feed industry, the FDA raised the permissible levels of selenium by a factor of three. Therefore, for the past six years, we've been able to buy feeds that contain 0.3 ppm selenium in complete feeds or 90 ppm selenium in trace mineral mixes. And supplement feeds had to be formulated so that sheep would receive a maximum of 0.7 mg of selenium per day.

In 1986, the old selenium levels were only one-third as high. To wit, until 1987, the FDA regulations stated that selenium levels could be no higher than 0.1 ppm in complete feeds and 30 ppm in trace mineral mixes for sheep, with a daily allowable maximum intake of 0.23 mg per head. These levels caused problems on farms and ranches, and everyone knew it. Selenium requirements for sheep hover around 0.1 ppm — but that number seems clear and tidy only on paper. On any farm or ranch, real sheep ignore their requirements and consume whatever they fancy. In practice, this means that some animals eat more than others, some soils are very selenium deficient or are extremely reluctant to release selenium which makes the resulting forages extremely deficient, some salt feeders are located in weird places, and sometimes, just sometimes, the salt runs out and the dreaded "oh-yes-the-salt-is-out-I'll-get-to-it-tomorrow" syndrome takes effect.

These situations may seem mighty bewildering to bureaucrats, but they were decidedly understood by feed suppliers, nutritionists, and veterinarians. In selenium-deficient areas of the country, prior to 1987, these professionals saw endless cases of white muscle disease and sudden death, even in flocks that were given selenium-containing feeds throughout pregnancy and lactation. To cope with this situation, their recommendation was simple: inject the lambs (and maybe the ewes) with a veterinary-prescription product containing selenium and vitamin E.

The 1987 revisions were like a breath of rational fresh air. The higher feed levels of selenium were permitted in response to the unavoidable variations of animal consumption, soil types, feeding practices, etc.

But not everyone was happy. A few organizations speculated that excess selenium from these higher levels *could potentially* escape into watersheds, contaminate fish runs, accumulate in the food chain, etc., etc. In short, they were concerned that sheep and other livestock *might* excrete forms of selenium into the environment that could *potentially* be dangerous. These groups petitioned the FDA to reconsider its 1987 ruling, which it did, and in 1993 the FDA found that, in fact, it agreed that it did not agree with its earlier agreeable findings. The FDA re-examined its data and concluded that, after all these years, not everything is known about selenium (an astute ruling). The computer models used by the FDA to predict environmental catastrophes cried out for coefficients, and the data to support those coefficients simply wasn't there. More research was necessary. Therefore, fellows, back to the drawing boards . . .

So here's a worry: where does this game plan end? An enterprising person could take that Federal Register entry and substitute nearly anything for the word "selenium" — like copper, or ammonium chloride, or limestone. The answer is always the same — for any substance, we don't know *everything* about its *every* metabolic form and derivative. We can never measure *everything*, let alone know enough to predict *all* the effects and residual levels under *every* conceivable situation. The bottom line is that, if the FDA changes or withdraws rulings based on a finding of incomplete knowledge, then the regulatory system becomes a poker game with everything wild — this is not a good game.

Two more interesting aspects of this charming situation:

First, in its long dissertation in the Federal Register, the FDA did not mention or document a single instance where excess selenium from agricultural feeding practices was actually found in the environment. It would seem that after six years of elevated selenium levels in feeds, if people were truly concerned about selenium runoff or environmental contamination, someone would have looked for it. Apparently no one did. Or if they did, they didn't find any.

Second, what about watershed contamination? Apparently, some fish are extremely susceptible to very low levels of selenium in their water, as low as 1 ppb (that's parts per *billion*). When livestock receive more selenium than they need, they excrete the excess via feces and urine. Fecal selenium is mostly in the forms of elemental selenium and metal selenides, which are not very water soluble. Urinary selenium is primarily methylated, which makes it more soluble (a logical adaptation, since mammalian urine is liquid). Excess feed selenium is excreted primarily in the feces, but selenium already in the body is released primarily in the urine. Here's the rub: If reducing the selenium levels in feeds forces livestock owners to increase their use of injectable selenium, how will the excess selenium be excreted? Probably in the urine. Which *could* put more water-soluble forms of selenium onto the ground, which *could* wash into the streams more easily than the fecal forms of selenium. So . . . the new (old) FDA ruling could actually *increase* the amount of water-soluble selenium in the environment. Makes sense, doesn't it?

But don't tell the FDA. They might conclude that more research is needed . . . to characterize the relationship between these excreted forms of selenium and the giant red spot on Jupiter.

First Published: December 1993

Author's Note: A few months after this article was published, the Federal Congress stepped in and ordered the FDA to cease and desist. At least until it gathered actual scientific evidence to support its concerns about an impending environmental catastrophe caused by the levels of selenium in livestock feeds. I'm pretty sure that this article did not make Congress intervene. But you can see the rest of the story in my article entitled "Kesterson."

Evolutionary Ruminations

Many folks spend their free time doing things like fishing or playing golf or dancing, but nutritionists spend their free time thinking about things nutritional. So as the result of some well-spent free time, I'll use this month's article to ruminate about two interesting nutritional items.

My first rumination deals with small numbers. Someday when you get the chance, look at a table listing the nutrient requirements for your animals. You can find these tables in many nutritional reference books — the NRC bulletins, the *SID Sheep Production Handbook*, the *Cow-Calf Management Guide*, etc. Pay particular attention to the mineral requirements.

One thing that you'll immediately notice is that some mineral requirements are expressed as a percentage of the diet — calcium, phosphorus, magnesium, potassium, sodium, chlorine, and sulfur. These elements are often called the *macro* minerals, because they are needed in large (macro) amounts.

Requirements for the other minerals are expressed in *ppm* (parts per million). Since these elements are only needed in small amounts, they are called the *micro* minerals. Look at these levels closely. Iron, manganese, and zinc are required in the range of 20–50 ppm. Copper is required at slightly lower levels, in the range of 7–13 ppm. But three minerals — cobalt, selenium, and iodine — are required in very, very tiny amounts — at only 0.10–0.30 ppm. This is intriguing.

Why? Because the three minerals that animals need in the smallest amounts are the same minerals that plants don't need at all.

How interesting. I don't think that this is an accident. In the broadest sense, evolution may be random, but on the molecular level, evolution is rigorous and unforgiving. It has forced our livestock to use rare resources in the most efficient way.

From a biochemical perspective, plant and animal cells must accomplish many of the same metabolic tasks and therefore contain similar metabolic systems. With only a few exceptions, plants require the same array of nutrients as animals. Most of this overlap occurs in the macrominerals and with some microminerals. It's true that plants also have their own special needs for elements that our livestock cannot use, like boron, which is involved in plant fiber metabolism, and molybdenum, which is used in plant nitrogen metabolism and nitrogen fixation. But what about the three minerals that plants *don't* need — iodine, cobalt, and selenium?

Well, animals use iodine almost exclusively in thyroid hormones. Since these hormones are critical for controlling metabolic rate and heat production, plants don't have much need for them or their iodine. Unless, of course, plants evolve into warm-blooded organisms — a concept I'd rather not contemplate.

The second mineral, cobalt, only occurs in animals as a component of vitamin B_{12} (properly called *cyanocobalamin*). All animals use vitamin B_{12} in amino acid metabolism and red blood cell formation, and ruminants especially use it in the liver for converting rumen fermentation products into glucose. But again, plants do not have red blood cells or rumens or even livers, and thus they have no need for either vitamin B_{12} or cobalt.

And selenium — a favorite topic of workshops and Internet discussion groups — is used by animals primarily in cellular anti-oxidant systems and also a few specialized enzymes (including some that work with iodine). Most plants contain no metabolic systems that require selenium. But remember the Periodic Table? Elements in the same column of this Table have similar chemical characteristics. Look at the placement of selenium. Directly above selenium, in the same column, is *sulfur*. Sulfur, of course, is an integral atom in two essential amino acids — methionine and cysteine — and thus plants require relatively large amounts of it. Since selenium chemically resembles sulfur, plant roots don't differentiate them very well and will absorb selenium from the soil along with sulfur. Plant cells happily incorporate this selenium into amino acids to form *selenomethionine* and *selenocysteine.* The plants insert these seleno-amino acids into various proteins in place of the regular sulfur-containing amino acids. Although those proteins may not function as well, the aberrant amino acids do no real harm because their levels are so low.

You might ask, "What about those famous selenium-accumulator plants — species of *Astragalus* and others — that can accumulate selenium to more than 1,000 ppm?" The exception actually proves the rule. Accumulator plants accumulate selenium in a specialized molecule called *Se-methylselenocysteine*, which does not occur in our standard forages. In terms of our discussion here, we can consider those accumulator plants as curious anomalies.

So let's engage in some wild speculation. Iodine, cobalt, and selenium show up in most forages because the absorptive roots allow them to come along for the ride. Although plants don't need these elements, these atoms generally do no harm, and they are absorbed in small amounts if they occur in the soil. Animals, on the other hand, require all three minerals for critical metabolic systems. And it all comes back to evolution. Evolution ruthlessly selected animals that could survive on low levels of these minerals. Animals that genetically required higher levels of these minerals tended to leave the gene pool early.

My second rumination is about cyanide. This topic is indeed nutritional, although cyanide is not exactly considered a nutrient.

You probably have heard about plants with the potential for cyanide toxicity — apple seeds and wild cherry and young sorghum-sudangrass. You may have also heard that very low levels of cyanide are usually not dangerous because animals can detoxify these molecules.

But have you ever considered *why* animals — livestock and birds and humans — have the metabolic ability to detoxify cyanide? Let's think about this . . .

When cyanide is absorbed across the gut wall, the blood carries it to the liver and other cells, where a specialized enzyme system converts cyanide into the non-toxic molecule *thiocyanate*, which the kidneys then dump into the urine for excretion. But enzyme systems are expensive. It costs energy and nutrients to manufacture all these enzymes, and evolution is notoriously frugal about overspending. So why do animals spend nutritional capital just to build extra metabolic equipment that they rarely use?

Because cyanide is not rare. It's actually a rather common molecule in the plant world. Plants use cyanide compounds as weapons in their ongoing biochemical warfare to stay alive. Cyanide discourages certain animals like slugs, weevils, and grasshoppers. Plant species that contain toxic levels of cyanide are well-known, but a surprising number of other plants can contain low levels of cyanide, including forages like white clover and common vetch and even some human foods like cassava and lima beans.

But cyanide is extremely toxic to animals. In animal cells, cyanide irreversibly binds to the critical enzyme *cytochrome oxidase*, which controls the last step in the extraction of energy from carbohydrates using oxygen. Cyanide is so devastating because it completely blocks the main energy-producing sequence in a cell. That's why animals who die of cyanide toxicity have blood that's bright cherry red — because their cells cannot use the oxygen in the blood, and thus the hemoglobin cannot get rid of it and stays bright red.

Again, let's think about nutritional evolution. Animals like to eat. If plants contained low levels of cyanide, and animals who ate those plants didn't have mechanisms for detoxifying cyanide, those animals wouldn't survive very long. This includes graziers, browsers, and species like us omnivorous humans who eat plants as well as animals. Eons of evolution therefore exerted relentless pressure. Genetic lines that developed the metabolic equipment for coping with low levels of cyanide did a better job of reproducing than genetic lines that lacked this equipment. Maybe not enough equipment to allow us to consume a meal of cyanide truffles, but enough to keep us alive when we encounter the occasional cyanide molecule.

Enough. It's time to stop ruminating about nutrition and spend some free time fishing.

First Published: December 2005
Author's Note: For everything there is a reason. A few years ago, researchers discovered that the human brain contains *opiate* receptors. As do the brains of other mammals. Why opiates? It's unlikely that our human ancestors lived in opium dens. But evolutionary logic is similar to our example with cyanide — nature's reasons must survive natural culling. These reasons became clearer when researchers subsequently discovered a new family of compounds called *endorphins*.

SECTION 3

Vitamins

Vitamins Through the Rumen

Everyone knows about vitamins, right? We need them, we buy them, we feed them, and then our animals become healthy and happy. Well, maybe but maybe not. Our animals are not humans. They're ruminants or alpacas or horses, and their peculiar digestive anatomies mean that we often worry about things that we *don't* need to worry about, and that we may miss things that we *should* worry about. So let's engage in some straight talk about feeding vitamins to livestock and see where it leads.

From my nutritional perspective, there are three fundamental categories of vitamins — fat-soluble vitamins, water-soluble vitamins, and vitamin C. The practical questions are relatively simple. Should we include these vitamins in diets? If so, when? If not, why not? I'll discuss these categories in reverse order.

First an important note: in this article I am referring primarily to *ruminants* — sheep, goats, cattle, deer, elk, giraffes, bison, musk-oxen, etc. These all have a rumen, complete with a few zillion rumen microbes that are quite capable of manufacturing certain compounds the host animal can later absorb into its own blood. I'm also talking about those rumen-like animals called *camelids* — alpacas, llamas, camels, and such — which chew cud and boast a rumen with said rumen attributes, but do not have cloven-hooves or a fore-stomach compartment called the *omasum*. Weird, but it works. And I'm also talking about *horses*, which most definitely *do not* have a rumen. But horses *do* have a huge large intestine which effectively plays the corresponding role of a rumen — it contains rumen bugs that manufacture useful compounds that the horse can absorb into its bloodstream.

Now on to vitamin C. Properly called *ascorbic acid*, vitamin — is the well-known antioxidant found in citrus fruits. It's also known as the scurvy vitamin — the reason why the British navy inventoried citrus fruits on their sailing ships, why British seamen are forever known as *limeys*. Vitamin — is actually a small molecule derived from glucose using the enzyme *L-gulonolactone oxidase* which most animals possess. Only a few species lack this enzyme — humans, some higher primates, some fruit-eating bats, guinea pigs, and the red-vented bulbul bird. An exquisitely select club to be sure — and they all require vitamin — in their diet. But ruminants are *not* in this club. Ruminants possess this critical enzyme and happily manufacture their own vitamin C. Therefore, we don't need to feed vitamin — to our livestock. Not unless you are trying to raise a herd of chimpanzees.

Next are the water-soluble vitamins. These are the famous B-vitamins. Whoa! I can almost hear folks exclaim, "You mean B-complex?" Well, without getting too complex, that terminology actually refers to a commercial injection that is a cocktail containing many different B-vitamin compounds. The reality is that there is no single "B-complex" vitamin. The B-vitamins are actually a group of eight unrelated compounds: thiamine, riboflavin, niacin, biotin, folic acid, B_6 (also known as pyridoxine), pantothenic acid, and B_{12}. All these compounds are soluble in water, and all are needed daily by our livestock.

Here's the good news: *rumen microbes can make all of these B-vitamins.* That is, during the fermentation process, the rumen bacteria synthesize these molecules, which then can be absorbed by the host animal and used for its own metabolism. Which means that, under normal conditions, *we don't need to add B-vitamins to ruminant diets.* Based on solid research conducted over the past 80 years, we think that a healthy population of rumen microbes makes enough B-vitamins to supply the needs of the host sheep or cow or musk-ox. This principle also applies to camelids, and, as far as we know, generally to horses.

The one exception, kind of, is vitamin B_{12}. This molecule is a bit of a nutritional oddity. Rumen bacteria are actually quite capable of making it, but they need a critical component first. This component is the element *cobalt*. The formal chemical name for vitamin B_{12} is *cobalamin*. The "cobal" indicates that the molecule contains cobalt, which means that the microbes require some cobalt to manufacture B_{12}. And in practice, this is exactly how we meet the ruminant requirements for vitamin B_{12} — we include cobalt in our trace mineral mixtures to provide it to the rumen microbes. We don't need to add pre-formed B_{12} to our ruminant diets or TM mixtures. In regions where plant cobalt levels are low, we can include 40 ppm cobalt or so in our TM mixtures and stop worrying about it. Interestingly, this is not the only nutritional strategy for supplying B_{12} to ruminants. In some countries, farmers effectively approach the B_{12} problem by applying cobalt directly to pastures in fertilizers or by using slow-release cobalt boluses, but these techniques are not practiced in North America.

I would add a practical caveat to ruminant B-vitamin nutrition — rumen microbes function well when they function well. Any syndrome that impairs rumen function, like acidosis, or any severe stress or disease that reduces feed intake may greatly alter microbial activity and thus reduce B-vitamin synthesis. Also, some toxic plants contain compounds that specifically target certain rumen vitamin pathways, like bracken fern *thiaminase* which may destroy rumen thiamine before it can be absorbed by the animal. Under normal, steady-state conditions, a healthy rumen cranks out B-vitamins without problem. But under other conditions, all bets are off: call a nutritionist.

So what's left? The fat-soluble vitamins: A, D, E, and K. These molecules are unrelated to each other, so I'll discuss them individually. With one exception, animals can't synthesize them. The exception is — you guessed it — vitamin D. When exposed to ultraviolet light, specialized enzymes in the skin will transform cholesterol into a vitamin D precursor called *cholecalciferol*. The blood then transports this compound to the liver and kidneys, where it is ultimately converted to the active form of vitamin D called *1, 25-dihydroxycholecalciferol*. A similar precursor can be found in bleached hay. The bottom line is that if our livestock roam outdoors in the sunlight for a reasonable period each day, they won't suffer from vitamin D deficiency. Of course, some animals such as orphan lambs, orphan calves, barn-housed horses, and confined dairy cows don't see much sunlight. For those situations, we definitely need to add vitamin D to their diets.

What about vitamin K? Usually getting sufficient vitamin K is not a problem because there's lots of it in green forage. Also, some rumen microbes can synthesize it in reasonable amounts. Either way, animals usually receive enough vitamin K to satisfy their requirements, so we don't need to worry about it. That is, unless our animals consume some really unusual feeds that contain antagonists of vitamin K, like moldy sweetclover hay (*Melilotus alba* and *M. officinalis*), certain sulfa drugs, or Warfarin® rat poison. But that is a story for another time.

For vitamin E . . . well, historically, we thought that vitamin E was used by livestock in ways similar to selenium, and that green forages contained enough vitamin E to satisfy most requirements. But we are learning that we have probably underestimated vitamin E requirements, especially for certain classes of animals, like high-producing animals, animals under stress, animals grazing mature forages, and confined animals fed high-grain diets. So now we are increasing the requirements and also routinely adding extra vitamin E to some diets and TM mixtures.

Finally, vitamin A. Fresh green forages usually contain lots of vitamin A (actually its precursor, the *carotenoids*). But deficiencies can occur in confined animals or animals grazing bleached, drought-stricken pastures. Also, hay that has been stored for a year or longer probably contains no vitamin A or vitamin E activity. These vitamins oxidize over time and lose their potency, even if the hay remains green. One good thing, however, is that unlike other vitamins, vitamin A is stored in the liver, and when animals come off an extended period of grazing green pasture, their livers contain 3+ months of vitamin A in reserve. In many cases, this reserve quietly gets them through a deficient period. But if they become deficient, we can always make sure that the TM mixture contains vitamin A, and as a last resort, inject livestock with a single megadose of vitamins A and D.

Vitamin A, however, is the only vitamin that we need to worry about toxicity, at least in theory. For example, if a 150 lb ewe requires 3,648 IU of vitamin A, a toxic dose would be more than 409,000 IU. This toxic level is generally hard to reach. But . . . but . . . recall that vitamin A is fat-soluble and accumulates in livers. Well, fish oils and seal livers contain lots of vitamin A. Who eats fish and seals? Polar bears. And sure enough, polar bear livers contain toxic amounts of vitamin A. The Eskimos knew this and avoided eating polar bear livers. So if your local feedstore advertises a sale price on ground polar bear liver, don't buy it.

First Published: December 2007

Author's Note: This is the most basic summary of vitamin nutrition for ruminants. In general, supplying vitamins to our livestock is easy, once you know the principles. Understanding those principles can save lots of unnecessary concern and necessary money.

The Vitamin that's More than a Vitamin

We hardly even think about Vitamin D anymore. It's one of those vitamins that we routinely add to feeds, an inscrutable listing of IUs on the feed tag. And as for our children, we simply irradiate the milk. And really, as a livestock producer, when was the last time you saw a case of rickets? Sure, nutritionists annually warn producers about rickets in animals housed indoors for months, as sometimes occurs with some horses and orphan lambs, but the lack of vitamin D is an anomaly, right? I thought so too, until recently, when I reviewed some new research about this vitamin. I found things that really made me sit up and think . . .

First, we should cover the basics about vitamin D.

We get this fat-soluble vitamin from two different sources — animals and plants. The animal source is a precursor molecule in skin epidermis called *7-dehydrocholesterol*. Ultraviolet light changes this molecule to *cholecalciferol* (*Vitamin* D_3), which is then carried by the bloodstream to the liver. In plants, a corresponding ultraviolet light process creates a different molecule called *ergocalciferol* (*Vitamin* D_2). This compound is ingested in the feed, absorbed through the gut wall, and then transported to the liver. In both cases, liver enzymes add a hydroxyl group to form *25-hydroxycholecalciferol* (25D). This 25D molecule then moves to the kidney, where another enzyme transforms it to *1,25-dihydroxycholecalciferol* (1,25D), which is the true active form of the vitamin D. It's this 1,25D molecule that acts as an on-off switch in cells throughout the body to regulate calcium balance, thus controlling a cascade of calcium-related processes such as calcium absorption from the digestive tract, bone formation, milk production, milk fever, and so on.

To continue our classic description, textbooks always list *the* main deficiency of Vitamin D as *rickets* — the bone deformation syndrome of children. Also *osteomalacia* which is rickets in adults. And also, to some extent, *osteoporosis*, a long-term problem of reduced bone density in older adults. Livestock can experience similar calcium-related bone disorders, as well as an additional syndrome of *milk fever* in cattle, which occurs when the onset of lactation drastically reduces blood calcium levels. A similar disorder in sheep, called *hypocalcemia*, occurs in late pregnancy or early lactation in twin-bearing ewes. All these syndromes, however, are still related to calcium balance.

We measure vitamin D in *International Units* (IU). One *micro*gram of D_3 equals 40 IU. For humans, the National Institutes of Health (NIH) recommends a daily dose of 400 IU for adults 51–70 years, slightly higher for elderly adults, and slightly lower for younger adults and children. For sheep and goats, the new NRC *Nutrient Requirements of Small Ruminants* recommends 5.6 IU/kg body weight during maintenance and early pregnancy. Which translates to 392 IU for a 154-pound ewe — remarkably similar to the human dose. (The NRC recommends higher doses for late pregnancy, lactation, and growth.) Let's remember that vitamin D has always been known as the *anti-rachitic factor*, and all these official NIH and NRC recommendations are designed to prevent rickets and other bone-related disorders, human or animal.

Livestock producers rely on sunshine to provide D_3 to their animals or on sun-cured hay to provide D_2, or they simply add D_3 to mineral mixtures or other supplements. In the human world, not many foods naturally contain high levels of vitamin D, so we irradiate milk to fortify it with vitamin D, or take pills containing D_3, or run around in the sunshine without sunscreen, or drink a tablespoon of cod liver oil every day (ugh!). Your grandmother may not have been a biochemist, but she was right about one thing — cod liver oil really *is* good for you. It's one of the few human foods that naturally contains lots of vitamin D_3.

Usually, that's the end of the story. But recent medical research has completely revised our understanding of vitamin D. From my perspective, this compound is far more critical than we ever dreamed. Here are some reasons:

For starters, even though it will always be known as a "vitamin," *Vitamin D is really more than a vitamin*. Typically, vitamins are small molecules that act as cofactors in larger molecules — vitamins become parts of large proteins like enzymes to make these enzymes work properly and catalyze the same reaction over and over again. For example, vitamin K fits like a jigsaw piece into certain blood-clotting enzymes to assist with their carboxylation reactions. Similarly, vitamin B_{12} is incorporated into three different enzymes to facilitate the transfer of hydrogen ions or methyl groups.

But Vitamin D is a horse of a very different color. Vitamin D is manufactured in one tissue (skin), is transported to two other organs for activation (liver and kidney), and then is transported to cells all over the body to regulate their actions. Uh, isn't this exactly the description of a *hormone*? Yes, vitamin

D is indeed a hormone. Its "vitamin" moniker is really just a historical accident. Vitamin D was discovered during the golden age of vitamin advances in the early 20th century. In fact, it was the fourth such molecule found — hence the name "D" (after A, B, and C. Get it? To confuse things further, only one B-vitamin — thiamine — had been identified by then).

Upon entering a cell, 1,25D is conveyed to the nucleus where it combines with a specialized *Vitamin D Receptor* (VDR) which then links with another molecule called the *Retinoid-X Receptor* (RXR). This VitaminVDR-RXR complex then binds directly to DNA at multiple sites adjacent to various target genes. This act of binding induces the target gene to activate or shut down, depending on the gene. Kind of like a toggle switch. We now know that 1,25D exerts some regulatory control over *at least 1,000 different genes.* And only a few of these genes are involved in calcium metabolism. The others are involved in some surprising things . . .

For example: two genes activated by 1,25D code for the production of two small peptides that have powerful antimicrobial action — *cathelicidin* and *defensin-beta-2.* In other words, the body can make its own natural antibiotics, which is a very good thing indeed.

Add to this knowledge an extraordinary discovery — that skin cells can create 1,25D from D_3 *by themselves.* In other words, these epidermal cells contain all the necessary enzymes to transform D_3 into 1,25D, and thus they avoid the requirement to transport D_3 to the liver and kidney first. Why is this important? Current speculation is that such a localized system allows these epidermal cells to respond to a potential infection threat by rapidly increasing the amount of antimicrobial compounds in the skin, which is certainly a useful trait for people trying to stay alive in a fearful microbial world.

Speaking of speculations . . . Tuberculosis has been a scourge of mankind since the pharaohs. But long before the discovery of antibiotics, doctors knew that people suffering from TB seemed to improve when they move to hot, dry places like Arizona or the Mediterranean. It was the effect of the dry climate, they assumed. But now we have an inkling of a real mechanism for TB control, and it's related to vitamin D. The higher level of UV light in these hot, dry places maximizes the endogenous production of D_3, which increases the production of those antimicrobial compounds, which seem to be at least partially effective against the TB bacteria.

Another gene regulated by 1,25D controls the level of *cytokines*, which are small molecules released by some immune cells during an inflammatory response. The binding of 1,25D to these genes reduces the amount of cytokines secreted by these cells, which seems to reduce the inflammatory response. Here's the gig: certain human autoimmune diseases are related to long-term exaggerated inflammatory responses of the immune system. Researchers are currently wondering if 1,25D or its analogues (chemical look-alikes) could be useful in treating or preventing autoimmune diseases like autoimmune diabetes, inflammatory bowel disease, and even multiple sclerosis

And then there is the potential link to cancer — currently a very active field of medical research. Scientists are generating a growing body of evidence that higher doses of vitamin D may reduce the risk of certain types of cancers. In the mechanistic metabolism world, researchers have demonstrated that 1,25D or similar compounds exert control on some genes that regulate cell growth. Also that these compounds reduce the growth of cells with damaged genes — kind of like putting a brake on cells that have the potential of unrestricted cancerous growth. This topic is not just restricted to laboratories. Last summer, the Canadian Cancer Society formally increased its Vitamin D recommendations for all adults to 1,000 IU supplementation during the fall and winter months. (And all year for higher-risk adults. Canada's high latitudes mean that it receives only low levels of UV light). This is a complex situation, of course, because exposure to too much sun is clearly a causative factor for some skin cancers.

But how do all these fundamental roles for vitamin D relate to livestock? Well, if we accept the premise that the current Vitamin D recommendations for livestock are based only on the avoidance of rickets, then we might consider that we may be missing an entire level of requirements. Who knows what levels of vitamin D are needed for optimal long-term health, not just for avoiding rickets?

Meanwhile, please excuse me. It's getting near my noon lunchtime. I think I'll go outside and sit in the sun awhile.

First Published: January 2008

Author's Note: This is an example where the medical research has outstripped the livestock research. The NRC vitamin D recommendations for livestock are still based on research that focused primarily on rickets and other calcium-related syndromes. Meanwhile, for my clients, I seriously scrutinize the vitamin D levels in their mineral mixtures and make adjustments as necessary. And personally, I make sure that I take at least 1400 IU of vitamin D daily.

A, B, C, D, E . . .

I've been thinking about a recent spate of scientific articles concerning the benefits of supplemental vitamin E on livestock performance and health, specifically in sheep, beef cattle, and dairy cattle — and even on the color of the steaks. But something here doesn't compute. We've known about vitamin E since the 1920s. So why the sudden renewed interest in it?

The reasons may be twofold: Firstly, because whenever we think of vitamin E, we also think of selenium, automatically, almost a knee-jerk reaction. We know that both prevent white muscle disease and both work together to prevent oxidation in cells. For example, think of the common injections of Bo-Se and Mu-Se. In our minds, vitamin E and selenium are as intertwined as bacon and eggs.

Historically most of the excitement has focused on selenium. Researchers, legislators, and feed manufacturers have all been intrigued by selenium's dual personality — a nutritional equivalent of Dr. Jekyll and Mr. Hyde: the good Dr. Jekyll with its prevention of white muscle disease, and the frightening Mr. Hyde with its alkali disease and Kesterson Reservoir. As we learn more about selenium's physiological roles — in retained placentas in cattle and immune function — its star continues to shine and attract research dollars. Vitamin E, on the other hand, has quietly played a background role, at least in the livestock world. This is an interesting anomaly, because we've long known that vitamin E will cure the poultry disease *encephalomalacia*, where selenium has no effect, and that vitamin E influences reproduction in many species — indisputable evidence that vitamin E is important in its own right, independent of selenium.

But this leads into the second reason for the resurgence of interest in vitamin E: its laboratory assay. Historically, the laboratory assay of vitamin E has always been tedious and difficult, especially compared to the selenium assay. The problem lies in the fundamental difference between minerals and organic molecules. Most mineral assays are relatively straightforward procedures of inorganic chemistry. A sample is exposed to high levels of energy, like heat or light. The minerals in the sample produce or reflect light at very specific wavelengths. To quantify selenium, we simply measure the presence and intensity of the wavelengths unique for selenium. Accurate selenium assays have been widely available for more than 50 years.

But vitamins are quite different. Vitamins are small organic molecules that often come in multiple forms. While selenium is easily described by its concentration (usually in mg/kg, the same as ppm), vitamin E is described by its *biological activity*, which is expressed in *International Units* (IU). One IU of vitamin E activity is universally defined as 1.0 mg of *dl*-alpha-tocopherol acetate, which is a synthetic compound created by industry.

The official name for vitamin E is "tocopherol" (pronounced "toe-ka-for-all." As in "free-for-all"). To make things even more confusing, tocopherol is not a single compound — it's a *family* of compounds. *Eight* different naturally-occurring compounds actually show vitamin E activity: four types of tocopherols (ingeniously named *alpha, beta, gamma,* and *delta*) and four types of closely-related tocotrienols (similarly named *alpha, beta, gamma*, and *delta*). Of these, alpha-tocopherol is the most potent; the other compounds show only 1%–40% as much vitamin E activity. Ironically, the universal standard for vitamin E — *dl*-alpha-tocopherol acetate — is a synthetic tocopherol that is not even found in natural feedstuffs.

Also, the *dl* in *dl*-alpha-tocopherol stands for the terms *dextro* and *levo*. Like many organic molecules, the tocopherols come in two forms that are mirror images of each other. Dextro means right-handed; levo means left-handed — referring to the way atoms are arranged around a non-symmetric carbon atom. The *dl* signifies a 50:50 mixture of two forms. Although the *d*-form is more than twice as potent as the *l*-form, manufacturers find it easier to produce the *dl*-mixture than the pure *d*-form.

I sometimes like to think that vitamin E is really nature's practical joke — designed just to prove that nature is not as simple as it sometimes appears. As in . . . once you think you have figured something out, you find that there's more to figure out. Vitamin E is like that. For any feedstuff, its vitamin E activity can reside in one, some, or all of the tocopherols and tocotrienols. For example, corn grain contains 35% of its vitamin E activity as alpha-tocopherol, 26% as beta-tocopherol, 26% as gamma-tocotrienol, and 13% as alpha-tocotrienol. Fresh perennial ryegrass, on the other hand, contains 100% of its vitamin E activity as alpha-tocopherol.

If that wasn't enough, the natural forms of vitamin E aren't even very stable — they have the distressing tendency to deteriorate over time. Oxygen will combine with them in a process called *oxidation*, which destroys vitamin E activity. Other destructive factors include heat, storage, low pH, plant maturity, ultraviolet light, heavy metal salts, and even the presence of unsaturated fats.

These fats are a special case. Over time, unsaturated fats oxidize and become rancid. Vitamin E prevents this oxidation — by substituting itself as the oxidized target. Therefore, high levels of unsaturated fats (like corn oil) can actually accelerate the loss of vitamin E in feeds.

So if you want to measure vitamin E, what *exactly* do you analyze? When? And how?

This is the crux of the problem. Because vitamin E is an organic molecule, you can't simply burn it like a mineral and then measure the resulting wavelengths. The original vitamin E assay developed by researchers was a long and daunting biological trial that traced the reproductive efficiency of rats. Technology made things slightly easier in the 1960s, with the use of various solvents and column chromatography techniques, but the assay was still slow and complex. Only recently however, the development of high-tech reverse-phase HPLC (high performance liquid chromatography) has made the vitamin E assay relatively easy. And now scientists can efficiently study the effects of vitamin E in all sorts of useful situations . . . like the health and performance of our livestock.

First Published: January 1998

Author's Note: As we learn more about Vitamin E, we are gaining a growing appreciation of its metabolic influence. Of course the popular press jumps on this bandwagon and tends to portray Vitamin E as a magical potion, a golden bullet that will solve all the world's health problems. Well, everyone is entitled to their own opinion. But beneath all that hype, the reality is that vitamin E *is* important and is probably required in greater amounts than we previously thought.

K as in Coagulation

This is a story of bureaucratic bungling, a Nobel Prize, three brilliant scientists, two countries, moldy hay, and rat poison. I am talking, of course, about the discovery of Vitamin K.

Before we start, I'll point out something — a very important something — about a uniquely American feature found in scientific papers from our land grant universities. The next time you read a scientific article in a research journal like the *Journal of Animal Science* or *Journal of Dairy Science*, look at the footnotes at the bottom of the first page. Among the usual items acknowledging the funding agencies and listing the authors' current addresses, look for a footnote with the words "Published with the approval of the Director of the [. . . state . . .] Experiment Station as Publication No. [. . . number . . .]." Although this type of footnote was more common years ago, you still occasionally see it today. I always thought it was just boilerplate fluff. Until now.

Back to the story. In the 1920s, farmers across the high plains were coming into their veterinary clinics carrying buckets of blood that wouldn't clot. They described a strange hemorrhagic syndrome in which their livestock, especially cattle, bled to death from the slightest wounds. No one knew the cause, but the main commonality between these farms was that the affected animals had been fed moldy hay made from a popular legume called sweetclover (*Melilotus alba* and *Melilotus officinalis*).

During this same period, poultry producers reported a strange bleeding disease in their chickens. These producers were beginning to follow the modern practice of raising chickens in wire cages, and they observed that some caged chickens suffered from a hemorrhagic syndrome that kind of resembled scurvy. Again, no one knew the cause.

Let's put these reports into a historical perspective. The early years of the 20th century was a period of explosive growth in scientific technology, especially in nutrition and biochemistry. Researchers developed powerful laboratory techniques for identifying toxins and other biochemical agents. It was during this period that scientists discovered the nutritional factors called *vitamins*, and vitamin research was all the rage.

These two bleeding syndromes, however, were puzzling, and researchers from many countries raced to discover solutions. In 1929, a prominent Danish biochemist named Henrik Dam published a paper that thoroughly described the bleeding syndrome in chicks. After additional research, Dam published another paper in 1934 in which he contended that no known vitamins were involved in this syndrome. Two Kansas researchers, Romayne Cribbett and John Correll, reached the same conclusion in their 1934 paper "On a Scurvy-like Disease in Chicks."

Meanwhile, at the University of California in Berkeley, S.F. Cook and K.G. Scott were conducting experiments with chickens and in 1935, they postulated that this bleeding syndrome was caused by some sort of unknown toxin in fishmeal (because feeding meat meal did not cause the problem). Their paper didn't make the fishing industry happy, but it did seem to point in a logical direction.

At the same time on that campus, a young scientist named Herman Almquist was also working on this bleeding problem in chicks. After earning a Ph.D. in organic chemistry, he had recently joined the Division of Poultry Husbandry in the College of Agriculture (yes, Berkeley had an agricultural school back then). In a series of brilliant experiments, he demonstrated that this bleeding syndrome involved a "factor" that was soluble in fat and that could be produced by bacterial growth in feedstuffs. He showed that he could prevent the bleeding syndrome by adding this factor back into purified diets, even diets containing fishmeal. Almquist concluded that this preventive factor which was somehow associated with the clotting process was a new, yet-undiscovered vitamin.

This was great investigative work, but then he tried to publish it. Remember that footnote I mentioned earlier? Well, the routine procedure for scientists on the Berkeley campus was, prior to sending their papers to research journals for publication, to submit their manuscripts to the Experiment Station Director's office for "approval."

But the university administrators had a serious problem with Almquist's paper — they didn't agree with his conclusions. Very strongly. They noted that senior scientists on their campus had found a different result (the possibility of a toxin) and also that respected scientists in other institutions had already concluded that no vitamins were involved in this bleeding syndrome. So these Berkeley administrators, in their wisdom, were afraid that Almquist's outlandish paper would embarrass their institution. They felt so strongly about it that they ordered Almquist to stop submitting research manuscripts until the matter was resolved within the university.

So . . . while Almquist fought to obtain approval from his university administrators, his landmark paper sat on a shelf waiting to be published in a scientific journal.

But the world did not wait for the University of California. In Denmark, Henrik Dam was actively researching the same syndrome, and in 1935, *he* published a paper in the journal *Nature* entitled "The Antihemorrhagic Vitamin of the Chick. Occurrence and Chemical Nature," with essentially the same results as Almquist. Although Dam and Almquist had both reached the same conclusions about this clotting factor, Dam published his paper first. And since the Germanic term for blood clotting is *Koagulation* spelled with a "K," Dam named this factor *Vitamin K.*

In 1943 Henrik Dam was awarded the Nobel Prize for his discovery of Vitamin K. Eventually, Herman Almquist won his argument with the university system and published his paper in *Nature* — a few months after Dam. But the damage was done. Almquist lost out on the Nobel Prize because he couldn't get his paper published first.

But that's not the end of my story. Remember that other hemorrhagic problem in cattle fed sweetclover hay? Well, Vitamin K is an integral part of the clotting process. Sweetclover naturally contains a compound called *coumarin.* Although coumarin is harmless, its derivative is not. When sweetclover is made into hay, and if that hay gets wet so that mold grows on it, the mold converts coumarin into *dicoumarol.* Dicoumarol interferes with the function of Vitamin K, and this interferes with the clotting process. Hence the bleeding syndrome in livestock fed moldy sweetclover hay.

After the discovery of Vitamin K, a Wisconsin scientist named Karl Link worked on this hemorrhagic disease of sweetclover and in 1941 successfully characterized dicoumarol and its anticoagulant properties. His laboratory then synthesized other similar compounds with anticoagulant properties, trying to find a commercial blood-thinning drug. Because of the financial profit potential of such a compound, the University of Wisconsin was very interested in his work, In fact, like many universities that administer the licenses and royalties from the commercial products of university research, the University of Wisconsin had set up a separate funding organization called the *Wisconsin Alumni Research Foundation (WARF).*

One compound that Karl Link synthesized from coumarin was particularly effective as an anticoagulant, far more powerful and dependable than dicoumarol. This compound was soon commercialized as a rat poison. In fact, it has become one of the most widely used rodenticides in the world, as well as a premier anticoagulant drug in human medicine. We know it as *Warfarin*®. But think for a moment about its name . . . Karl Link had cleverly combined the acronym of his university foundation with the name of the base compound — *WARF* plus coum*arin.*

So there we have it. University bureaucrats delay the publication of a breakthrough paper on Vitamin K because they wish to avoid institutional embarrassment, and in doing so they deprive a fine scientist of a Nobel Prize. Years later, after working on a compound linked to Vitamin K, another fine scientist in a different state names a rat poison after his university's funding organization.

Life is filled with little ironies.

First Published: December 2006

Author's Note: For years, I have told this story to my livestock nutrition classes. Everyone, of course, has heard of Warfarin, but no one had ever considered where the word came from, or why the letter "K" was associated with this vitamin. But if you look up a 1975 review article by H.J. Almquist in the *American Journal of Clinical Nutrition*, you can see where he refers to his initial research with chicks. His words, of course, are gracious and generous, but this is a case where it's even more interesting to read between the lines.

Nardoo & Opisthotonos

Opisthotonos, polyneuritis, beriberi, polioencephalomalacia, bracken, and nardoo. What do these words all have in common? Thiamine.

Thiamine?!! Is thiamine deficiency a problem in ruminants? Not usually. Despite some enthusiastic claims to the contrary, sheep and cattle usually get enough thiamine from their digestive tract. In the rumen, some microbes routinely manufacture thiamine for their own private metabolism. This is very convenient for our livestock, because after each bug dies, its thiamine can then be absorbed across the digestive tract and used by the host animal. Convenient, simple, and effective. Except when something goes awry. Then we see a problem — classically a downer animal showing neural symptoms which dramatically respond to thiamine injections. Veterinarians usually call this disease *polioencephalomalacia*, which is a very fancy term for thiamine deficiency in ruminants. We need to note here that polioencephalomalacia has *NOTHING* to do with human polio, which is an infectious viral disease controlled by vaccinations. Unfortunately, polioencephalomalacia in animals is also sometimes called polio, but the animal syndrome is a nutritional deficiency, not an infectious disease.

So why discuss thiamine now? Because it's spring, the beginning of our growing season. And there are a couple of plants we should know about.

Some chemistry first: thiamine is a funny-looking molecule, kind of like an unbalanced dumbbell, with two dissimilar rings at each end linked together by a short chain of carbon atoms. Sometimes, things come along that split this chain — like certain types of sulfur compounds which are often found in molasses. Also, an entire family of enzymes called *thiaminases* make their living by cleaving that carbon linkage and destroying thiamine molecules. Unfortunately for our livestock, some plants specialize in producing thiaminase enzymes. If our sheep and cattle eat enough of these plants . . . goodbye thiamine, hello polioencephalomalacia.

Two of these plants are fairly common: horsetail and bracken fern. Most of you have seen horsetail (*Equisetum arvense*), a short, non-flowering, hollow-stemmed plant that resembles a horse's tail (at least it did to someone). If you rub it between your fingers, you'll find that it's surprisingly rough to the touch — it has a high silica content. In fact, I used to use it instead of steel wool to scrub pots during camping trips. It often grows on wet ground and is rather unpalatable to livestock. Horsetail contains all sorts of toxins, including thiaminase.

Bracken fern (*Pteridium aquilinum*) is even more common — it's the *classic* fern plant in meadows and open woodlands across North America. It often grows in very thick stands. People sometimes collect the young bracken shoots, called "fiddleheads," as a gourmet delicacy, which is unfortunate because these shoots contain carcinogenic (cancer-causing) compounds.

Livestock don't usually eat horsetail or bracken because those plants are not very palatable. However, in the early spring these plants often come up quite heavily at a time when other green feed is short, and some animals, especially exploratory lambs and calves, may munch them in spite of our wishes. Also, during the summer these plants may end up in the hay, particularly in heavily infested fields. I've seen worrisome stands of these plants here in fields all around Douglas County. The hay-drying process apparently does not inactivate the thiaminases. Therefore, if large amounts of bracken fern or horsetail are in the pasture or hay (particularly bales that come from certain areas of a field), at 20% or higher, the hay may contain enough thiaminase to cause a problem. Horses are particularly susceptible to thiamine problems if they consume this type of hay for an extended period of time (enough time to deplete enough thiamine from enough cells). Thiamine deficiency can also show up in other scenarios — high-grain diets, certain feed additives, high sulfur levels, and of course the "I-haven't-a-clue" situation that occurs when you least expect it. Basically, there's a lot we *don't* know about the production and destruction of thiamine in sheep and cattle.

However, a lot *is* known about thiamine in human nutrition. Thiamine is the vitamin we all learned about in high school — you know, the cause of *beriberi* in humans and *polyneuritis* in chickens. You may remember your lessons about the first discovery of vitamins — about Dr. Christiaan Eijkman observing beriberi in people fed polished rice in Jakarta (then called "Batavia") in Indonesia during the late 1800's. Eijkman conducted his experiments with chickens, and when he fed them brown rice or rice polishings, the problem went away. Thus, thiamine was the first B-vitamin discovered (hence the name "Vitamin B1." It's a good thing that we didn't discover 235 B-vitamins — that naming system would have become a bit cumbersome). But thiamine's discovery did not mean that beriberi disappeared quickly. A lot of education had to take place first. In fact, as late as 1947 in the Philippines, thiamine deficiency was the second leading cause of death after tuberculosis.

Which brings us, of course, to *nardoo*. This is a strange and unlucky story: Nardoo (*Marsilea drummondii*) is an Australian fern that grows near streams. In 1861 two explorers, Robert Burke and William Wills, undertook an epic journey across some of the most distant regions of the Australian outback. They trekked from central Australia up to the northern edge at the Gulf of Carpenteria and then back again. But as they neared their base camp during the return trip, they ran out of provisions and tragically could not locate the supply caches left for them by their support team. Starving, they subsisted on the only green plants in the area, nardoo, which at that time of year grew abundantly along the banks of Cooper Creek. Unfortunately, nardoo contains quite a bit of thiaminase — *at levels 100 times greater than bracken fern*. For thousands of years, the local aborigines had successfully used this plant for food, but they made it safe by first soaking it in water before eating it. The aborigines tried to give advice, but the hungry explorers insisted on using a different method to prepare the nardoo, and they continued to eat their fill of it. Their appetites remained good until the end, but they lost energy, could not travel quickly enough to find help, and eventually died, probably of thiamine deficiency.

And finally, there's *opisthotonos*. What a word! Try this word out on your veterinarian — she may be so impressed that perhaps she'll give you a discount on your next invoice. In any case, opisthotonos means stargazing. Not the romantic kind of teenage viewing of the heavens from the front seat of a Rambler, but the frightening, muscle-clamped, strained position of an animal with its head thrown back over its shoulder. It's almost too painful to look at. Opisthotonos is a common symptom of thiamine deficiency in livestock. But don't automatically assume the animal has polioencephalomalacia . . . many other problems can cause this dramatic symptom, including listeriosis, lead poisoning, and rabies. So if you see opisthotonos, handle the animal carefully . . . and check the ration.

First Published: March 1994

Author's Note: Nutrition has often played a significant role in history, and not just the lack of it because of siege warfare or naval blockades. Books and the Internet are filled with fascinating stories like the one of ill-fated Burke and Wills expedition that ended tragically on Cooper's Creek.

History versus Today

Over the past couple of months, I've written two articles about vitamins, which got me thinking. So please indulge me — I'm going to engage in a bit of rumination . . .

Let's imagine, for a moment, that we are nutritional scientists conducting research on critical factors in feeds. We are particularly interested in something called "Factor X." We don't know much about Factor X — which is why we named it "X" and are doing research on it. But we do know this: if we feed animals diets that don't contain Factor X, they suffer from something called the *Horrid Syndrome.*

Now, because this is a magazine article and not a scientific laboratory, we can peak under the hood. Unlike the real world, where scientists must work in the dark to shed light on their topics, as magazine readers *we* can access *all* the information about Factor X, even before the scientists learn about it. And in the amazingly complex world of biochemical systems, it turns out that Factor X is intimately involved with a whole array of metabolic systems. We can see, from our perspective as all-knowing readers, that Factor X clearly interacts with metabolic Pathways A, B, C, D, E and F.

Now let's switch back to our scientific laboratory where we are just researchers rather than all-knowing magazine readers. Our scientific experiments are just beginning to shed some light about Factor X, and it appears that Factor X may be an essential dietary component for preventing the Horrid Syndrome. Over time, we conduct more experiments. Finally, after years of painstaking laboratory work, we discover compelling evidence that Factor X is intimately linked to Pathway A. Additional experiments successfully allow us to outline the detailed biochemistry of Pathway A, including a method to synthesize it. Now we can see the details of how Factor X influences this pathway. And there's more — we learn that animals will suffer from the Horrid Syndrome when Pathway A doesn't work properly. We also learn that we can prevent the Horrid Syndrome by adding purified Factor X to the diet. Our goal is in finally sight.

We feverishly direct our research to determine the optimum dietary levels of Factor X to support the proper functioning of Pathway A and prevent the Horrid Syndrome.

After more years and more research, the general scientific community fully accepts the conclusion that Factor X is a required nutrient. Official government documents dutifully list the recommended levels of Factor X that will prevent the Horrid Syndrome. Our laboratory continues its work on Factor X to flesh out additional details of its chemistry and structure. The Horrid Syndrome has been conquered. People are overjoyed. We receive worldwide accolades; we are granted tenure at a prestigious university; there is talk of a Nobel Prize.

But in the back of our minds, we hear a little nagging voice. What about those *other* metabolic Pathways — B, C, D, E, and F? During the halcyon days of our research, when we were racing to solve the problem of the Horrid Syndrome and Pathway A, we didn't spend any time investigating those other pathways. In fact, we had no idea that Factor X was even remotely involved with them. We concentrated our efforts on the Horrid Syndrome, and preventing it became our overriding goal. Based on this research, the published dietary requirements for Factor X were designed specifically to prevent the Horrid Syndrome.

That little nagging voice continues: what if those other pathways are also critical? What if our dietary recommendations are *too low* for those other pathways, but *their* deficiency symptoms are more subtle than the Horrid Syndrome? *What if we missed something important?*

Okay, let's end our little fairy tale. The Horrid Syndrome is just a figment of my overactive writer's imagination. It doesn't really exist.

Or does it?

Let's change the words: for Factor X let's substitute "*vitamin D*" and for the Horrid Syndrome, let's substitute "*rickets.*"

Here is the real-world, non-fairy tale scoop: rickets became a widespread scourge in the late 1700s at the beginning of the industrial revolution, when people left their farms and moved into smog-filled cities to work in factories. By the mid-1800s, doctors learned that cod liver oil and sunbathing could both prevent rickets, but no one knew why. During the heyday of vitamin discoveries in the first decades of the 20th century, scientists worked hard to identify the mysterious nutritional factor in cod liver oil with anti-rachitic properties. *Anti-rachitic* is the scientific term for something that prevents rickets. By the mid-1930s, researchers had discovered two forms of vitamin D: ergocalciferol (D_2) and cholecalciferol (D_3). By the late 1960s, scientists had worked out the complex metabolic pathway that showed how vitamin D influenced calcium absorption and thus bone health and density. Nutritionists pegged the dietary levels of vitamin D to prevent rickets in children and its analogue in adults, osteomalacia. Similarly, in the livestock world, animal nutritionists set vitamin D requirements to prevent rickets and other calcium-related diseases in livestock.

Those dietary requirements have not changed in many years. In fact, the authors of the 2007 NRC reference book *Nutritional Requirements of Small Ruminants* retained the 1981 vitamin D levels because they couldn't find enough data to justify a change. In other words, what was good enough in 1981 is good enough today.

But recent research in human medicine has overtaken our earlier knowledge about vitamin D. We've recently learned that vitamin D has a fundamental role in controlling gene expression, which affects such diverse pathways as immune function, cell growth, inflammatory response, and the formation of localized antimicrobial compounds in the skin. None of these pathways were understood or even considered when scientists developed the original dietary recommendations for vitamin D. Researchers in 1981 did not have our modern technologies to manipulate genes, and they did not know of the existence of thousands of control compounds which we now have characterized in exquisite detail. But things are changing. In human medicine, which may move a tad faster than the livestock industry, the Canadian Cancer Society recently upped its dietary vitamin D recommendations by 150%.

And here's a different wrinkle in our fairy tale: for Factor X and the Horrid Syndrome, let's change the words to "*selenium*" and "*white muscle disease*."

For years, selenium research in livestock was designed to prevent the obvious syndromes of selenium deficiency, such as white muscle disease, exudative diathesis in poultry, mulberry heart disease in hogs, etc. In the U.S. after much regulatory wrangling (because the window between selenium requirements and toxicity is rather narrow, and also because some 1940s research linked selenium to cancer), the federal regulations of 1987 currently limit selenium to 0.3 ppm in a total diet for livestock, with similar tight caps on selenium levels in supplements and free-choice mineral mixtures.

But today's scientists have technologies their predecessors never dreamed. We can now detect things in parts per *trillion* (that's one millionth of a ppm), and we can manipulate genes and cells in ways that were considered science fiction when I was a kid. Recent research has shown that selenium is involved in many other things besides the obvious syndromes, such as (1) immune function and possibly cancer prevention, (2) selenium-based enzymes control thyroid hormone and thus can influence basal metabolic rate, and (3) some molecules containing selenium sequester heavy metals which may influence their toxicities, especially in a contaminated world. But given this new research, have we revised our selenium recommendations for livestock and humans? Are we *certain* that our current selenium recommendations are sufficient for *these* functions? Have we carried out extensive nutritional studies to find out? In all cases, the answer is no.

And I've discussed only two nutrients. What about all the other vitamins and minerals?

I suspect that for some of these nutrients, we may be underestimating their true nutritional requirements. We've been spectacularly successful at eliminating their Horrid Syndromes, but what about their potential B, C, D, E, and F pathways? We need to uncover a lot more of the total biochemical picture, and we may need to adjust our dietary recommendations. Yes, it's harder to research subtle metabolic pathways, and it's even harder to quantify nutrient requirements that properly support those pathways. But "harder" does not mean that higher requirements don't exist.

All we need, of course, is a bazillion dollars for research and a centrifuge the size of Kansas. No problem.

First Published: February 2008

Author's Note: Am I recommending that everyone run out immediately and double or triple the dietary levels of all their vitamins and minerals "just to be safe?" Of course not. Such a recommendation is not justified without good scientific evidence and a prudent analysis of the field situation. Increasing nutrient levels will-nilly may do more harm than good, with unforeseen interactions and all sorts of interesting nutritional problems. But this only underscores the need for more research. We need to explore different syndromes, different pathways — it's a little like thinking outside of the box.

SECTION 4

Requirements, Ration-Balancing, Foods

Ration-Balancing is Like a Square Dance

All join hands and circle left;
Circle to the left, eight hands around!

A good livestock ration is a lot like a square dance. It has a clean structure, no-nonsense efficiency, and no matter how complicated, it serves its purpose precisely and well. But if the design is flawed — well, you know what a bewildering place a dance floor can be . . .

First let's analyze a typical square dance. To newcomers, the whirling dance may seem like chaos, but there really *is* logic to those movements. Trust me. Every square dance is built from only a few repeating elements. Each dance contains a basic figure, which includes simple or complex moves. Dancers perform this figure with each other in the square, and when this basic figure ends, the dancers may have come back to their original partner or they may have temporarily changed partners. In addition to the basic figure, every good square dance also contains a chorus figure, called a *break.* This break figure is done between repetitions of the basic figure, over and over again, much like the chorus of a song. Sometimes dancers perform the break with their original partner, sometimes they don't. But when all the dust settles at the end of the dance — for all the thousands of square dances that have been written, some with incredibly complicated figures — the dancers always end up back at their home positions with their original partners, and all is well.

Like the square dance, every good ration is built from only a few basic elements: a base forage, a source of vitamins and minerals, an energy supplement, a protein supplement, water, and a carrier for drugs or other feed additives. A nutritionist combines these basic elements in various ways to provide enough energy, protein, vitamins, minerals, and water to meet our animals' needs. Also, like a square dance, a good ration should use its elements economically, with tightly controlled precision. No duplication, no excess nutrients, no unnecessary costs. Excess and duplication cause nutritional problems and hurt the pocketbook.

Yet even with these restrictions, there are nearly endless combinations of ration elements. For example, mineral mixes usually include trace minerals, except when they don't. Some mineral mixes may also include feed additives and vitamins, but sometimes they don't. Some supplements contain high levels of both energy and protein, but some supplements are purposefully deficient in protein. Sometimes supplements aren't even needed. Most rations contain only one forage, but at certain times of the year, good rations may contain two or more different forages.

Allemande left with your left hand . . .
Back to your partner with a right and left grand.
Hand over hand go 'round that ring,
Meet your partner and promenade!

Let's examine a ration for a simple situation: ewes and lambs on a lush ryegrass-clover pasture in western Oregon in mid-April. The land is as green as Ireland, and the clover understory is so thick that we're afraid that our animals may bloat.

Our ration contains an obvious base forage: the pasture. Let's assume, for a minute, that this pasture has enough energy and protein to meet our ewes' needs. The second component of our ration is the block of trace mineralized salt that we've put out on the pasture. It contains sodium chloride (white salt) and a complete spectrum of trace minerals. The third ration component is the water in the creek. It contains . . . well . . . water, and possibly some dissolved trace minerals but hopefully no runoff herbicides or nitrates.

If our ewes need extra energy — for example, to increase milk supply in the ewes raising twins — what supplement should we choose? How about a convenient protein block which may also contain additional minerals? I don't think so. Our base forage in April already contains more than 20% protein, which is enough for these ewes, and we are already feeding minerals. What about cull peas? No. Peas also contain too much protein. A liquid protein lick? No, we already have enough nitrogen in the forage, and a protein lick would be a very expensive way of supplying molasses. What about some third cutting alfalfa hay? No, the alfalfa would be a *second* base forage, which is redundant. Do you think sheep will consume hay while grazing lush pasture? Also, alfalfa doesn't contain enough energy to do the job.

An answer: Follow the basic principle of KISS — *Keep It Short and Simple* (that's one definition). Nutritionally, the simplest and most elegant option would be to choose a single source of energy, like corn or barley or oats.

Swing that gent,
And swing that girl,
Everybody swing,
And everybody whirl!

Now the dance moves smoothly, with rhythm.

Why even discuss this topic? Because we sometimes follow feeding strategies that contain duplicate ration elements and then wonder why things go wrong. For example, let's say that we feed a supplement containing grains, salt, and some (but not all) minerals. Our sheep and cattle will avidly eat that supplement, of course — but will they also consume the trace mineral block in the pasture, at least at the consumption levels we expect? Not a chance. Because the white salt in the supplement satisfies our animals before they ever get to the trace mineral block in the pasture. What if we depended on that trace mineral block to supply a critical dose of selenium or magnesium or a feed additive like Bovatec®?

The dance moves on. In late May, our Oregon forages change significantly as they rapidly approach maturity. Our simple energy supplement of corn was good in April but it would be a poor choice in May. The maturing base forage has become low in protein, so therefore the supplement must supply both energy *and* protein. We'll need to add cull peas to our corn in a 50:50 mixture to bring the entire supplement up to 16% crude protein.

The dances get more and more intricate . . .

Bow to your partner, corner too,
Thank the band —
That's it, you're through!

First Published: February 1995
Author's Note: A little artistry is good. Maybe a bit unexpected, but good. In another part of my life, I call square dances and contra dances — and we learn, as callers, the beauty of lucid, concise instructions. When our instructions are inexact or ambiguous or redundant, the feedback from the dance floor is nearly instantaneous, and the confusion is painfully obvious.

The Nutritional Value of a Truffle

Let's talk about the nutritional value of a truffle.

After all, this is February. A time of saving lambs, doing chores, and the sweet smell of hay in the barn. And Valentine's Day. Sweets for the sweet. Which brings us to the subject of truffles. Have you ever thought about how many calories are actually *in* one of those delicious, innocent-looking morsels?

Don't laugh — this is actually quite practical. Let's think of a truffle as a *complete feed* — kind of like a single day's ration of silage or hay. How would we solve the very typical problem of estimating this ration's nutritional value? The principles (rules) for this exercise are exactly the same as for any other ration. We are just applying those principles to a rather unique ration.

First a clarification: There are truffles and there are truffles. For the purpose of this discussion, I'm talking about the glossy chocolate candy bombs that people drool over in fancy candy "shoppes," not the black tuber-like fungus that hogs dig up in the woods.

So what is my special Valentine's Day feed? For this exercise, let's stick with the basics: a standard chocolate truffle from the Euphoria Chocolate Company in Eugene, Oregon. This is an approved feed for humans, although perhaps its intake should be prudently limited. Let's consider that one truffle equals our entire ration for the day. As for any other ration, we first must consider its weight. If we assume a 100% consumption (a reasonable assumption), then this weight equals the daily intake. Let's say this particular truffle weighs 1.6 oz (45 grams in scientific terms).

The basic question is this: How much energy and protein is in our typical truffle? Well, I am *not* going to grind up this perfectly good specimen for nutritional analysis, even in the name of science. Instead, I'll look up its nutritional value in a reference table. Hmmm — this presents a slight problem. A truffle is clearly not a typical feedstuff. It is not listed in the *SID Sheep Production Handbook*, or in Morrison's classic *Feeds and Feeding*, or in any other standard livestock text. But because a truffle is a mixed feed, I can try a different strategy. I can *derive* its nutritional value by looking up the individual ingredients and then combining those numbers for a total estimate.

So what's in a truffle? (Read: what's in a complete feed?) Most truffles don't come with feed tags, or warnings, for that matter. However, I contacted the feed supplier (the chocolate shop). After some energetic coaxing, I was able to extract its basic recipe from the feedstore manager (the chocolatier). Their standard chocolate truffle contains 0.4 oz (11.3 g) of heavy cream and 1.2 oz (34.0 g) of chocolate. Now we're getting somewhere.

But we still need some additional numbers. Specifically, we need to know the dry matter percentage of each ingredient; the percentages of fat, protein, and carbohydrates in the dry matter; and also the digestible energy content of the dry matter. Since fat contains approximately 225% more energy than carbohydrates, even a little bit of fat can make a big difference.

There are reference tables for these values. For example, I have a reference chart from the field of human nutrition that lists heavy cream at 58% water, which translates to 42% dry matter. This dry matter is composed of 88.0% fat, 5.0% crude protein, 7.0% carbohydrates, plus some vitamins and minerals. This table also lists the gross energy value of heavy cream at 8.2 Calories per gram of dry matter.

Now for the chocolate (do we *really* want to know?). There are many types of chocolate, but let's assume it's a standard milk chocolate. My trusty reference table lists it at 99% dry matter. This dry matter contains 32.5% fat, 7.2% crude protein, 57.7% carbohydrates, and for those who care, also some vitamins and minerals. The gross energy value of this milk chocolate is 5.2 Calories per gram of dry matter. Still with me?

All we have to do now is combine these values proportionally to arrive at a fair estimate of our truffle's nutritional value. Without boring you with the details, suffice it to say that this represents a typical situation for balancing rations. Our 45-gram truffle contains 11.3 grams of heavy cream, which equals 4.7 grams of dry matter (42% of 11.3 = 4.7), and 34.0 grams of chocolate, which equals 33.7 grams of dry matter (99% of 34 = 33.7). Then we add 4.7 plus 33.7 to arrive at 38.4 grams of dry matter for our entire truffle. Similarly, we add up the specific amounts of each component (fat, protein, and carbohydrates), divide by 38.4, and convert to a percentage. Thus, we calculate that our truffle's dry matter contains 39.4% fat, 6.9% crude protein, and 51.4% carbohydrates. By similarly adding up the energy values, we see that our 1.6 oz truffle contains approximately 215 Calories of gross energy. Bon appétit!

What about fiber? Well, I think we can safely assume that a chocolate truffle does not contain significant levels of fiber. Unfortunately, this means that its gross energy is probably fully digestible.

I must add something for completeness: truffles may also contain various secondary metabolites — compounds that plants produce in small amounts for a variety of reasons. For example, chocolate contains caffeine. This is fascinating, but perhaps better left for a different discussion.

So, there we have it. Our complete ration of one truffle is exceedingly high in energy and rather low in protein. But let's think about this for a moment. These calculations and assumptions are *exactly* the same as you would use for estimating the nutritional value of any mixed feed. Your ingredients, of course, would be somewhat different.

Happy Valentine's Day! In case you've now lost your appetite for truffles, you may instead consider including them in a supplemental ration for growing lambs. Remember, however, that truffles are low in protein and therefore should not be fed to lactating ewes. (Even the thought of trying to feed truffles to 200-pound Suffolks gives me pause).

Meanwhile, like all good scientists, I must conclude that this topic requires further research. Replication is needed, as well as a larger sample size. I clearly must visit the Euphoria Chocolate Company again for some additional samples. Also, I have my eye on a potential follow-up project: the Double Chocolate Truffle . . .

First Published: February 1994

Author's Note: Yes, this was an actual classroom exercise. I first did this exercise in a *Livestock Feeding* course for ranchers here in Oregon. After their initial exclamations of disbelief, the ranchers went through this exercise with smiles. Then we ate our samples — more smiles. I did refrain, however, from adding anything else into the ration, like olives.

New Year, New Thinking

It's a new year — a great time for thinking outside the box. In this case, the box is our common problem of balancing diets for our livestock. I want to introduce a new way of looking at it.

Whenever we try to balance a diet, what is the first thing we do? We find a reference book, look up the animal's expected level of dry matter intake (DMI), and dutifully use that number to govern our calculations. In essence, we tacitly accept the assumption that the book's reference value accurately describes the size of our animal's feed "package." Then we develop our nutritional calculations so that the sum of the feed ingredients fits tidily into that package.

But this approach involves two major problems. The first one, as any nutritionist will tell you, is that the largest variation in our nutritional calculations is the amount of DMI, and that a reference book value for DMI is, at best, an educated guess.

The second problem is that the diet's composition itself will influence the level of intake, and therefore our package size can change based on the ingredients in the package. In other words, the composition of a diet can change the intake of that diet, even though we use the original intake number to calculate the composition of that diet. This is particularly true for ruminants, where diet composition greatly influences the rumen environment. And the livestock that we are trying to feed are ruminants.

Even when we reluctantly acknowledge these two problems, we generally disregard them anyway and plod on with our calculators and ration-balancing software. But if we really want to understand our diets and better predict their effects on our animals, maybe we should question our assumptions.

In fact, someone has. Doug Hogue, professor emeritus of animal nutrition at Cornell University, has come up with a rather unusual approach to balancing diets, which he calls *The Dugway Feed Company Nutrient Plan*. Different, is, well, different. Hang on tight — this is going to be an interesting ride. The Dugway system includes some concepts that may be new to you.

Concept #1: The nutritional energy value known as *TDN* is really a "score" — a single pooled number designed to represent the amount of energy available in a feedstuff. Other common nutritional terms, such as *Digestible Dry Matter, Digestible Organic Matter,* and the Energy System *(Digestible Energy, Metabolizable Energy, Net Energy)* are all just variations of this score. Although modern diet formulation models use complex computer-derived scores, these values are still originally based on chemically-derived scores. But the energy reference value (score) of any feedstuff was initially determined by laboratory digestion trials in which that feedstuff was offered *at maintenance levels of intake* — without regard to the other components of the ration and also without regard to the effects of higher levels of intake (on fermentation time, rate of passage, digestibility, etc.). So if these scores are inherently inexact, why use them at all?

Concept #2: Feedstuffs containing high levels of quickly-fermentable material like starches and sugars can profoundly change the rumen environment and its fermentation patterns, which can cause metabolic disturbances. This not only impacts the original feedstuffs but also everything else in the rumen.

For example, if I add molasses to a diet, I am adding a relatively large amount of soluble carbohydrates. Rumen bacteria ferment these molecules quickly and release lots of acids, which reduces rumen pH to less than 6.0. In turn, this lower pH suppresses the population of other species of bacteria that ferment cellulose, thus reducing fiber fermentation of *any* other feedstuffs in the rumen that contain cellulose. This reduces the level of intake of that diet. Although this situation may not greatly affect high-grain diets containing little cellulose, it definitely affects the digestibility of forage-based diets which contains lots of cellulose. In formal nutrition classes, we routinely call this problem an *Associative Effect*, but in the field, a good diet formulation system should account for it.

So rather than building a diet in the traditional manner — namely by calculating the amounts of protein and energy contributed by each feedstuff, adding them up, and carefully manipulating the amounts of each feedstuff to fit into a pre-defined intake package, the Dugway Nutrient Plan identifies diets by their *percentages* of protein and fiber while limiting the energy levels to a target proportion. In short, this system more or less ignores levels of DMI and takes into account how feedstuffs affect rumen fermentation.

Here are the details: we are all familiar with dietary crude protein levels — 12%, 16%, etc. — no problem there. But for dietary fiber levels, the Dugway system attempts to optimize the amount of *fermentable fiber* in the diet, which means calculating the amount of *indigestible fiber.*

Whoa! A little bit of explanation here would be nice . . .

Modern laboratories report the total amount of fiber in a feedstuff as *NDF* (Neutral Detergent Fiber). An important maxim in the Dugway system is that NDF is nutritionally composed of two parts: (1) the digestible (fermentable) portion, and (2) the indigestible portion. Unfortunately, the NDF number doesn't indicate how *much* is digestible or indigestible.

But the Dugway system solves this problem in a practical way. Although we can't easily analyze a feedstuff for its amount of fermentable fiber (*FNDF* = Fermentable NDF), we can *derive* this value mathematically by identifying the amount of *non-fermentable fiber* (*INDF* = Indigestible NDF) and subtracting that number from the total. This INDF calculation is easy because we can use that old hoary standard value of TDN listed in many reference books.

We generally equate TDN with digestibility. Look at any nutritional textbook, and it will formally define TDN as the sum of the digestible portions of protein, fat (ether extract multiplied by 2.25 to account for the extra energy in fat), crude fiber, and something called "nitrogen-free extract." Without going into details here, let's accept the important concept that TDN represents the digestible portion of a feedstuff.

So wouldn't this mean that if we subtract the TDN value from 100, we would get the *indigestible* portion of the feed? Unfortunately, not quite. Although that number would indeed be the amount of material that comes out the back end, so to speak, some of that material is derived from the animal's metabolic processes and not directly from the feed — waste products like bacterial debris, digestive enzymes, gut cells sloughed off from the inside of the intestinal wall, etc. We call this endogenous material the *Metabolic Fecal Loss* (MFL). This MFL can represent 10–15% of the feed, so we need to correct the TDN value for it. Let's assume that sheep MFL is 12%, and let's use this number as our correction factor.

Now we are ready. We calculate the amount of INDF using the formula INDF = 100 – TDN – 12. Because NDF is composed of INDF plus FNDF, once we know INDF, we can easily derive FNDF by subtraction.

Here's an example: fresh early bloom *alfalfa* with a TDN value of 60% and an NDF value of 40%. We can find these numbers in a reference table or forage report. Based on these values, this hay's INDF is 28% (= 100 – 60 – 12) and therefore its FNDF is 12% (= 40 – 28). Thus, the fermentable fiber portion of this alfalfa is only 12%. Why is this important?

Well, let's take another feedstuff: fresh mid-bloom *orchardgrass* with the same TDN (60%) and an NDF value of 64%. Using the same formula, we see that this orchardgrass has the same INDF level of 28% (= 100 – 60 – 12), but its FNDF is 36%, *three times higher than the alfalfa.*

With such wide differences in fermentable fiber, these two feeds will have very different effects on rumen function, dry matter intake, and possible metabolic disturbances, not to mention their responses to pelleting.

A bit complex? Yes. But now let's relax for a bit. Next month, we'll use these concepts to fashion diets and reach some surprising conclusions that we can use in the field.

First Published: January 2006
Author's Note: There is nothing else in ruminant nutrition like this Dugway System. Does this system answer *all* of nutritional questions? Of course not. But by setting out a different intellectual framework for formulating diets, the Dugway System is a very good place to start asking questions and devising research trials.

Diets Outside the Box

I know you've waited breathlessly, so let's not waste any time. We'll pick up on last month's discussion about formulating diets — that is, formulating diets using the unconventional concepts of the Dugway Feed Company Nutrient Plan. You've heard about thinking "outside the box." Well, for this discussion, *we'll* be thinking outside the box.

Last month I introduced the notions of TDN scores, metabolic fecal losses, fermentable fiber, and indigestible NDF. It's easy to get lost in the terminology, so let's step back and look at a broad conceptual framework. This way, the details will hang together and make more sense.

The big picture is that four main concepts profoundly influence ruminant diets: (1) things that *do* ferment in the rumen, (2) things that *don't* ferment in the rumen, (3) metabolic disturbances, and (4) variations of dry matter intake. The Dugway Nutrient Plan uses these four concepts to formulate diets. Let's discuss them.

[An important note: Diets, of course, contain other nutrients like protein (which includes the entire array of rumen degradable and non-degradable fractions), fats, minerals, and vitamins. Although these are all nutritionally important, they inevitably comprise less than 20% of nearly any diet. For this month's discussion, we are really interested in the other 80% of the diet — and how the four concepts listed above affect it.]

Concept #1: Things that *do* ferment in the rumen fall into two major types: (a) *Non-Structural Carbohydrates (NSC),* which are mostly sugars and starches. These molecules ferment relatively quickly in the rumen, sometimes very quickly. (b) *Fermentable Neutral Detergent Fiber (FNDF),* which are types of fiber that rumen microbes colonize and ferment slowly, like cellulose and hemicellulose.

Concept #2: Things that *don't* ferment in the rumen include lignin, cutin, Maillard Products from heat-damaged hay and silage, fibers impregnated with silica (common in rice straw and many range plants), and some others.

Although these may not be chemically related, they all are rather inert as they pass through the rumen. They provide no energy to the animal or to the rumen microbes. Together, they all fit under the umbrella that we'll call *Indigestible Neutral Detergent Fiber (INDF).*

Concept #3: *Metabolic disturbances* — a critical concept, and one reason this Nutrient Plan is so different from anything else. Other methods of formulating diets assume that everything in the rumen works well all the time. But in reality, things often go nutritionally awry in the rumen or with the animal, especially when grain is fed, and these metabolic disturbances will greatly influence how a diet affects the animal and how the animal consumes the diet.

The main metabolic disturbance that concerns us is *acidosis*, also called *lactic acidosis* or *grain overload* or *founder*. Whatever we call it, acidosis is the syndrome where relatively large amounts of acid pour from the rumen into the blood and lower the blood pH. Every farmer knows about *acute* acidosis — the veterinary emergency when sheep or cattle get into the grain bin. But on a day-to-day basis, *subacute* acidosis can occur routinely with diets containing high levels of NSC. The sugars and starches ferment quickly in the rumen and reduce the rumen pH, which allows lactic acid-producing bacteria to thrive. Because lactic acid is a stronger acid than other rumen acids, it has a greater impact on blood pH. The resulting subacute acidosis causes animals to back off feed, even slightly. Think of the nutritional irony here: We feed grain to our animals to increase production, but the high NSC levels in that grain tend to reduce feed intake, which reduces the amount of nutrients the animals actually receive.

Concept #4: And finally, the *variation of dry matter intake* (DMI). Hmmm. Although all reference tables list values for DMI, appearances are deceiving. Those DMI values are just educated guesses in a world of wild variation. Many things influence DMI, including diet composition. We just noted how subacute acidosis will reduce intake. High NSC levels have another effect beyond subacute acidosis. High NSC levels lower rumen pH, which reduces the rate of fiber fermentation, which slows down the rate at which fiber leaves the rumen, which translates to increased "fill," which depresses DMI.

A few years ago, Doug Hogue at Cornell University conducted a simple nutrition trial in which he observed that Finn-Dorset ewes rearing triplets in early lactation consumed feed at 6.8% of their body weight, while also *gaining weight* at 0.55 lb/day during that same period. The 1985 Sheep NRC reference tables list that ewes in early lactation rearing twins will consume feed at only 4% of their body weight and will also *lose weight* at 0.13 lb/day during this period. How can we reconcile these numbers? Here's how — the ewes in that trial told us that we are not good at predicting DMI. And by the way, that ewe diet contained a high percentage of soyhulls, which are high in FNDF.

Which brings us back to the Dugway Nutrient Plan and how it formulates diets.

Every feedstuff is evaluated for NSC, FNDF, and INDF — these are general energy categories that compose 75–80% of any diet. Diets are then formulated on *percentages* of these categories. (The percentage of protein is fixed at 12% or 16% or whatever, to meet the animal's requirements). The percentage of each category is set to provide sufficient nutrients to the animal, obtain a good intake, and maintain a healthy rumen. The Dugway Nutrient Plan doesn't worry about assuming a fixed value for DMI and then trying to fit *amounts* of nutrients into that DMI package. The level of DMI will take care of itself.

The percentages of these categories are then adjusted for different levels of animal production. Diets for lower levels of animal production would contain a higher percentage of INDF and a lower percentage of NSC. As animal production increases (daily gain, milk yield, number of lambs, etc.), the percentage of INDF in the diet goes down, the percentage of NSC goes up, but very importantly, the percentage of FNDF either remains constant or declines very slowly.

FNDF is critical. FNDF provides energy to the animal, but unlike NSC, it does not cause a severe drop in rumen pH and therefore doesn't cause acidosis or reduce fiber digestion. By maintaining a high level of fermentable fiber (FNDF), the Dugway Nutrient Plan avoids acidosis, supports good rumen function, and provides an appropriate level of nutrients to the animal.

And under this system, some feedstuffs really shine — namely, feedstuffs that contain high levels of *both* TDN *and* FNDF. Last month, I explained how to calculate FNDF. Being a modern technology guy, I built a simple spreadsheet to do the calculations for me. I evaluated an array of common feedstuffs and came up with some interesting results. Feedstuffs with high levels of FNDF are soyhulls (56%) and beet pulp (40%). In contrast, corn and barley had FNDF levels of only 8% and 15%, respectively, which means that most of their energy is in the form of starch, which is not surprising.

But one comparison really stood out — oats versus soyhulls. Both feedstuffs list their TDN at 77%, but oats contain only 21% FNDF while soyhulls contain a whopping 56% FNDF. Which feedstuff would support better rumen function? If you scan recent research articles, which feedstuff is currently attracting the interest of nutritional scientists?

Interesting. But now we need to get back to reality. If you go to the library or search the Internet, I'm afraid that you won't find much written about the Dugway Nutritional Plan. There are only a couple of informal handouts from some Cornell seminars. There are no reference tables or textbooks or computer programs that describe the Dugway Nutrient Plan. Not yet. But remember that this is very new stuff, outside the box stuff. You might want to keep your eyes peeled. Like acorns and mighty oaks — give this plan some time to grow.

First Published: February 2006
Author's Note: One of these days I'll get around to creating a computer program to calculate diets using the Dugway System. Just as soon as I finish this book.

A New Take on Intake

When we devise a ration for sheep or cattle, we always first ask the question "how much will the animals eat?" Then we decide on a likely level of intake and build a ration based on that decision. We're not too surprised when the animals actually eat that amount of feed. But . . . the diet formulation was based on our estimate of the intake, and the intake was based on the formulation of the diet. Aren't we inside a logical box?

Let's climb out of that box for a moment . . .

Recently at Cornell University, Doug Hogue and some undergraduate students conducted a modest study with ewes raising triplets. Instead of guessing what the ewes would eat, they let the ewes tell them. The ewes were quite eloquent. Now we must question our own assumptions.

First, a little background on nutritional requirements: The 1985 National Research Council reference book *Nutrient Requirements of Sheep* (NRC — *the* reference source for diet formulation) lists the requirements only for ewes raising singles or twins, not triplets. We can, however, start with the listed numbers and then use our best judgment to adjust them for triplet lambs. In brief, the NRC nutrient requirements for a 144-pound ewe raising twins during early lactation are 3.85 lb of TDN and 0.91 lb of crude protein. The NRC tables also list the dry matter intake (DMI) at 6.0 lb, which equates to 4.2% of body weight. The underlying assumption of those requirements is to maintain a ewe's weight throughout her production cycle except during early lactation, when she would experience a negative energy balance. During peak lactation, the NRC expects twin-rearing and singlerearing ewes *to lose* 0.13 lb/day and 0.06 lb/day, respectively. We would, of course, expect triplet-rearing ewes to lose even more weight than that.

Based on these NRC tables, many experienced sheep professionals suggest the following as a rule of thumb for feeding ewes in early lactation: Give the ewe one pound of 16% grain for each lamb that she is rearing, plus all the good alfalfa hay she can eat. A 144-pound ewe rearing triplets, for example, would receive a daily ration of 3 lb of grain plus 4.0–4.5 lb of alfalfa hay (assuming a DMI of 6.5 lb = 4.5% body weight).

But Doug and his students reversed this logic. Rather than feeding a limited amount of grain and allowing unlimited access to hay, they fed a limited amount of hay and allowed unlimited access to grain. And they found that, not only did those ewes *not* die of acidosis, but their performance . . . well, read on . . .

The trial consisted of fourteen ewes all rearing triplets (Finn-Dorset crossbred ewes averaging 144 lb). The ewes were fed a severely limited amount of hay at only 3.3 lb/day but they were allowed to consume all the 16% grain supplement they wanted. The trial lasted for 41 days, beginning a few days after lambing — i.e., during the peak period of lactation. The lambs were sired by good black-faced bucks, so we know that those lambs had a pretty good genetic potential for growth. The lambs did not have access to a creep feed. The hay was just an average quality grass-legume hay, and the grain pellet was a 16% commercial supplement. For the purposes of calculating DMI, let's assume that the hay and grain both contained 90% dry matter.

The results: The ewes consumed 3.3 lb of hay *and 7.6 lb of pellets each day* — a DMI of 9.8 lb/day (after adjusting for DM percentage), which equals *6.8% of body weight*. During those 41 days, the lambs *each* gained 0.71 lb/day and — hold on to your hats — *the ewes also gained 0.55 lb/day.* And this was during early lactation, when the NRC expects ewes to *lose* body weight. Instead, those ewes supported 2.1 lb of total lamb growth each day while simultaneously adding over a half pound to their own weight. Not bad.

You might be wondering about the grain pellet. Well, the pellet was *not* a special formulation from the depths of Cornell's laboratories. It was simply a commercial 16% high energy lamb pellet, right off the feed store shelf. It contained mostly grain, with 15% forage as fiber source, some Bovatec® to control coccidiosis, 2% limestone to balance the calcium-phosphorus ratio, and 0.5% trace mineral salt. In other words, a fairly reasonable and routine formulation. Its fiber component, however, was primarily soy hulls. Soy hulls are high in pectin, which is a type of fiber that digests quite rapidly in the rumen but still retains the fermentation characteristics of other types of fiber. This may have helped, but it certainly wasn't the whole picture.

Now turn on your calculator. If that "average" hay contained 13% protein "as fed" — a reasonable assumption — then correcting for percent dry matter gives a crude protein value of 14.4% on a dry matter basis. That kind of hay would probably contain 60% TDN. Similarly, the 16% high-energy supplement (on a dry matter basis) would contain approximately 17.7% crude protein and 85% TDN. Applying these values to a DMI of 9.8 lb means that those ewes consumed 7.59 lb of TDN and 1.64 lb of protein, which are 97% and 80% higher, respectively, than the NRC requirements for twin-rearing ewes. Even if we allow an extra cushion for the higher requirements of triplet lactation, it's obvious that those ewes ate quite a bit, maybe 50% or more than their requirements, *as the NRC has defined those requirements*. But did the ewes get fat? No, they just reared triplets successfully while gaining weight at the same time.

So are the NRC requirements wrong? Probably not. The committee of scientists originally derived those requirements from lots of careful, solid experimental evidence. However, the underlying assumption of those requirements — i.e., that ewes cannot maintain body condition during early lactation — is patently wrong. Those fourteen ewes told us that. Apparently, those ewes had not read the NRC book.

It's a good thing, too, because the NRC book does not list the nutritional requirements for ewes in early lactation rearing triplets where the ewes gain 0.5 lb per day *while* providing enough milk for each of their lambs to gain 0.7 lb per day.

Perhaps we should think about reevaluating the nutritional requirements for lactating ewes, or at least reevaluating our strategies for feeding ewes during early lactation. And also reevaluating our assumptions about what a ewe can really accomplish. A 300% lamb crop without any orphan lambs or loss of body condition — that's a wonderful goal for any flock of ewes. Especially when we know how to feed them.

First Published: February 1997

Author's Note: This was only a small study, but it's actually earthshaking work. I derived many of the numbers in this article from the 1985 NRC. The new 2007 Small Ruminant NRC indeed lists nutrient requirements for ewes raising triplets, but its requirements for a triplet-rearing 144-pound ewe in early lactation expects a DMI of only 3.3% body weight, a daily weight loss of 0.09 lb, and a diet containing only 66% TDN. The assumptions from the previous editions haven't changed. But they should.

Ewes Off the Scale

The *Nutrient Requirements of Sheep, Sixth Revised Edition, 1985* is one of those remarkable documents that everyone relies on to balance rations, almost without question. Published by the *National Research Council* (NRC), its reference tables are concise and comprehensive. These tables neatly list the daily requirements for energy, protein, minerals, and vitamins for sheep managed in the typical North American production systems. But two words are operative here: "typical" and "1985." There has been a world of change between 1985 and today, and many sheep producers currently manage their flocks in ways that the NRC authors never dreamed about. Let's examine three new situations where the old requirements may not apply.

But before we begin, I'd like to take a moment to review the basic structure of the NRC tables. The NRC requirements are based on the premise that ewes must receive enough nutrients to maintain body tissue and then receive additional nutrients to support production. The NRC assumes that a ewe will lamb once each year. It then divides the ewe's year into five distinct phases based on its production physiology: maintenance (15 weeks), early-mid gestation (15), late gestation (6), early lactation (8), and late lactation (8). A total of 52 weeks. The number of lambs influences the actual nutritional requirements. Late gestation ewes carrying a 130–150% lamb crop require less nutrients than ewes carrying a 180–225% lamb crop. As you'd expect, early lactation ewes suckling twins have the highest nutrient requirements. Early lactation ewes suckling singles are listed with the same requirements as late lactation ewes suckling twins. And late lactation ewes suckling singles are listed with the same requirements as late gestation ewes carrying 130–150% lamb crop. Ewe lambs, of course, have their own separate table of requirements.

The premise of maintaining body condition has one exception: early lactation — when ewes are allowed to lose body condition. The NRC requirements anticipate a small weight loss for ewes rearing singles and a greater weight loss for ewes rearing twins. During early lactation, ewes are expected to draw down their stored fats to support peak milk production. Ewes then regain this weight during late lactation. But if they can't regain this weight during late lactation, they still have the luxury cushion of another 15 weeks of maintenance to build up their body condition.

Accelerated lambing systems, however, have changed things indeed. Two of the most popular systems are the *STAR System* (five lambings in three years) and the *3-in-2 System* (three lambings in two years). Both systems share a critical nutritional characteristic — they reduce the ewe's production cycle from five periods to three. How? By eliminating the periods of late lactation and maintenance.

This makes sense, really. Regardless of anyone's wishes, gestation length is always fixed at 21 weeks (early-mid period plus late period). Ewes on accelerated systems, however, suckle lambs *only* during their early lactation period. They don't have a late lactation period — it disappears because lambs are weaned at 60 days or earlier. The maintenance period also disappears because the ram is introduced immediately after weaning. If everything goes according to plan, the ram and ewe quickly initiate another gestation period, and thus the cycle begins again.

Let's think about this. Accelerated lambing systems eliminate an important time cushion built into the normal NRC production cycle. This means that accelerated ewes who lose body condition after lambing must regain it very quickly — either at the end of their short lactation period or during the first part of their next gestation. Not much time for mistakes. Alternatively, these ewes can be fed to maintain body condition during lactation. Either way, accelerated systems push ewes harder than the assumptions built into the NRC tables, and the lactation and gestation rations for these ewes should reflect their higher requirements.

But even for systems with ewes on an annual lambing cycle, what about *ewes trying to rear triplets or quads*? During the last few weeks in late gestation, all those fetuses are gaining 60–70% of their fetal weight, and then during early lactation, those same ewes are trying to pump out milk like a high-producing dairy cow. The nutritional requirements for those ewes are certainly much higher than the requirements for twin-bearing ewes. But how much higher, precisely? The NRC is silent about these girls. So in practice, we feed them lots of grain and everything else in the barn, and we also provide a creep feed to the lambs. Well, if these ewes are on an annual cycle, at least they have that luxurious maintenance period of 15 weeks. That's a good time for them to fly to the Bahamas, lie in a hammock, catch some rays, and regain their body condition after being underfed during their production period.

And finally there are the *specialized breeds of dairy sheep* — such as East Friesian, Lacaune, and Awassi. These are ewes specifically bred for high milk production, in the true dairy industry sense. Their milk is used for the manufacture of sheep cheese and other dairy products. During the last few years, lots of commercial producers have purchased East Friesian ewes, or crossbred ewes containing East Friesian genetics, or composite breeds built from dairy breed genetics, like the Rideau Arcott. Unlike their dairy brethren, these commercial producers are *not* interested in constructing milking parlors or selling cheese, but they *are* interested in the extra milk and the potential for larger, faster-growing lambs, especially from multiple births.

But the 1985 NRC has an underlying problem: its table of nutritional requirements assume a standard milk production curve with a certain height and shape. The genetics of dairy sheep annihilate this assumption. Not only does the milk production of these ewes peak at a much higher level than non-dairy breeds, but that level remains higher throughout lactation. The milk production curve of dairy breeds is more *persistent* — i.e., it does not decline as quickly during lactation as in other breeds.

So what does this imply for sheep nutritional requirements? The genetics of dairy sheep breeds opens up a plethora of questions:

- How much extra feed is necessary to support milk production, especially during early lactation? Higher milk production requires more nutrients than for other breeds, or else the ewes will lose too much body condition.
- What nutritional level will maintain ewe body condition during lactation or at least allow ewes to minimize its loss?
- When does the early lactation period *really* end, or for that matter, when does the late lactation period end? Extending the length of early lactation may impact our decisions about when to wean the lambs and how to manage the risk of mastitis, and also how we define the genetic maternal effects on lamb gain (i.e., when does lamb growth begin to reflect its own genetic abilities for postweaning gain rather than its mother's milk supply?).

Hey, asking questions is easy; it's finding answers that is the real challenge. All we need is a couple of billion dollars for research. But I suspect that as soon a new edition of the NRC is published, someone will come along with something even newer, like a production system that uses dairy breed genetics in a crossbred ewe to raise triplets and quads in an accelerated lambing calendar, using management intensive grazing of high-fertility, high-producing forages. No problem. A supplement of taco chips will probably do the trick. Check the next NRC.

First Published: April 2000

Author's Note: Well, the next sheep NRC was indeed published in 2007 with the title *Nutrient Requirements of Small Ruminants.* This new NRC modified the categories of the ewe production cycle and also listed nutrient requirements for ewes rearing triplets. It didn't do everything I suggested — but then again, I would have been very surprised if it did.

The New Sheep NRC

In January 2007, the National Research Council (NRC) finally published the revision of its primo nutritional reference book for the sheep industry: the NRC nutrient requirements of sheep. People have been waiting for this document for a long time, but some folks are going to be surprised. This NRC book is more than just a simple upgrade of the slim volume published in 1985. In fact, it's no longer just a sheep book — it's been expanded to include goats, alpacas, llamas, and deer. The new title reflects this change: *Nutrient Requirements of Small Ruminants.*

This is a landmark book, so let's settle down, pour ourselves a cup or two of coffee, and spend the next couple of articles examining it — how it has changed from the previous edition, how can we use it effectively, is it worth the price, and does it have any problems or deficiencies?

First, a little about the organization that published it. The name "NRC" has an almost mythical ring, as if the NRC was a distant realm of all-knowing scientists who periodically hand down wondrous reference books. Actually, the real story is much more down-to-earth. The parent organization of the NRC is the *National Academy of Sciences* (now officially known as the *National Academies*). This is a private American institution funded by all sorts of grants. One of its divisions is the NRC, which consists of many subgroups, including one called the *Board of Agriculture and Natural Resources.* This Board, in turn, created a temporary working group called the *Committee on Nutrient Requirements of Small Ruminants* for the sole purpose of developing this reference book. This committee was made up of nine scientists from various institutions in the United States, Australia, and Mexico. They began meeting in mid-2004 and took slightly more than two years to complete the book.

The committee's mandate was simple but profound: to review all relevant scientific literature to date, evaluate the advancements since the previous NRC edition, and compile the most current recommendations for the nutrient requirements of these animals.

My role? I was one of the official reviewers of the document. Although I was not a member of the core writing committee, as a reviewer I had an extended opportunity to see this book before it became a book. It was my task to wade through the original draft document with all its imperfections. Now I can look at the final product with a wide perspective.

But before leaping into the new edition, let's take a quick overview of the previous editions for all these species, so we can better understand the features and changes in this one.

The previous Sheep NRC book was published in 1985 under the title *Nutrient Requirements of Sheep, Sixth Revised Edition*. A total of 99 pages, it included an index (7 pages), a list of references (14 pages), and extensive tables of feed composition (22 pages). The main chapters covered all the nutrients — energy, protein, minerals, vitamins, and water — with compact technical descriptions of basic nutritional concepts. Its chapter of nutritional disorders covered only four syndromes: overeating disease (enterotoxemia), polioencephalomalacia, ketosis (pregnancy toxemia), and urinary calculi. The 1985 edition also included a chapter on specialized sheep nutrition topics, such as flushing, early weaning, artificial rearing, creep feeds, poisonous plants, and range sheep. All of the chapters tended to be short — they were designed as dense summaries with just enough information to guide interested readers to more comprehensive books for longer explanations. (And this book did not contain a single reference to a website. Do you remember or can you imagine living in a world without the Internet?)

The core of the 1985 edition, of course, was its reference tables of nutrient requirements. The 1985 NRC listed requirements for (1) mature ewes weighing 50–90 kg raising either singles or twins, (2) ewe lambs, either pregnant or lactating, and (3) various categories of replacement and finishing lambs. The 1985 NRC divided a ewe's production year into six periods: maintenance, flushing, early-mid gestation, late gestation, early lactation, and late lactation. Daily nutrient requirements were listed either as total amounts or as percentages of the diet.

One of the basic assumptions underlying these nutrient requirements was that a mature ewe would maintain a steady-state, medium body condition throughout the year, with the annual *net* change in her body weight caused primarily by wool growth. During the year, her actual body weight would temporarily fluctuate due to gestation and lactation, but these changes were essentially add-ons to a steady-state animal.

Although most folks probably never read the 1985 Sheep NRC, those reference tables affected every producer because the tables were the basis of nearly all university and feed company sheep publications for the past twenty years.

In contrast to this sheep NRC, the nutrient requirements for goats were a bit on the thin side. The only previous Goat NRC publication was the 1981 volume entitled *Nutrient Requirements of Goats: Angora, Dairy, and Meat Goats in Temperate and Tropical Countries.* This rather sparse volume of 91 pages was filled mostly with tables of feed composition and descriptions of goat browse. It contained just three pages of nutrient requirements and only eight pages describing the nutrients. In fairness, it also included short sections on metabolic disorders and suggested rations, but it is clear that in 1981, not much was known about goat nutrition.

One thing we should understand is that an NRC "Nutrient Requirements" book is really a backwards-looking document. It reflects the nutritional state-of-the-art *in the year just prior to its publication.* The 1985 Sheep NRC was actually written in 1983 and 1984, which means that its reference tables reflect nutritional research published prior to 1984. The older Goat NRC book relied on research published prior to 1980 — *over a quarter century ago.* Think about how many major nutritional advances we have seen in the last 25 years — in fiber analysis, microbial yield, degradable protein, and computer models, just to name a few.

Let's complete our species review; what about alpacas, llamas, and deer? Alpacas and llamas are popularly known as *South American Camelids,* although the NRC scientists seem to prefer the phrase *New World Camelids,* Whichever name you use, these animals are nutritionally quite similar to ruminants — their 3-compartment stomach includes a rumen, and they chew their cud. Deer, however, are full-fledged, card-carrying ruminants. But for years, all these species have wandered in nutritional limbo. The animals, of course, didn't care — they ate their rations and grazed quite contentedly without the benefit of user guides or nutritional manuals. But their owners, and the supporting cast of feed companies, nutritionists, Extension agents, and veterinarians, were all operating in the dark. No one in North America had systematically reviewed all the scientific literature of these species or published tables of their nutrient requirements. Until now.

Now let's turn our attention to the new 2007 NRC publication: *Nutrient Requirements of Small Ruminants.*

Pick one up, and aside from the new title, two things that you notice immediately are its *size and weight*: This is a large hardcover book of *362 pages.* Whoa! Don't drop this book on your toes. You'll also notice its *price* — currently listed at $116 plus shipping at the National Academies Press website (http://www.nap.edu). We sigh wistfully for the days when the old NRC editions were small volumes that cost less than $30.

Now that you have a copy in your hand, open it up and turn to the reference tables beginning on page 246 and see . . . oops. It seems that we're out of space this month. Okay, please come back next month. We'll begin with some new things that are particularly interesting . . .

First Published: March 2007

Author's Note: This article is just the beginning. I've written four in this series — and they just barely scratched the surface of this NRC volume. Hang on for the ride.

Twenty-Five Pages of Numbers

The new Small Ruminant NRC is a big book with lots of features. My background as a university professor suggests that I have a tendency to, ahem, pontificate — that is, to say, that I tend to go on and on and on and on. But I will gallantly refrain from such grandiloquent oratory, even though, frankly, parts of this edition almost beg for it. Instead, I'll be succinct. This month, I'll focus on the real heart of this book — the reference tables listing the nutrient requirements for sheep. I'll use a broad brush to describe the big things, so you'll know where to look when you try to balance a sheep diet.

The 1985 Sheep NRC edition listed nutrient requirements on just seven pages. In contrast, in the new Small Ruminant NRC, the reference tables of sheep nutrient requirements now take up *25 pages*, plus another 23 pages of reference tables for feedstuff composition. Whew! And the difference is not a larger font. The tables in this new edition definitely contain more items and more categories.

I promised to be succinct, but here is an interesting aside that caught my eye: this section of the book is called "Nutrient Requirements of Sheep Tables." I kid you not. At first I thought this was a new scientific breakthrough. I didn't know that sheep tables *had* nutrient requirements. Were the authors referring to specialized ovine turning cradles or just routine shearing tables? The wooden tables in my shop don't seem to have nutrient requirements. But hey, what do I know?

Anyway, back to the numbers. Whenever you balance an animal's diet, you must first look up its nutrient requirements — which are based on the type of animal you're trying to feed, its stage of production, level of production, body weight, and occasionally some other minor factors. For the broad animal categories, this NRC lists requirements for mature ewes and rams, replacement ewes, and growing/finishing lambs.

Let's discuss mature ewes first. Reference table 15-1 divides the annual production cycle of a ewe into logical periods, similar to previous editions, but some details are critically different. The new production periods are maintenance, breeding, early gestation, late gestation, early lactation, mid-lactation, and late lactation. The word *breeding* here means *flushing*, with the nutrient requirements during this period increased 10% over maintenance. The major change in the production cycle, however, is that lactation is now split into *three* periods (early, mid, and late) rather than just two (early and late). The new NRC defines early lactation as the first 6 weeks after lambing, mid-lactation as weeks 7–12, and late lactation as beyond week 12.

Another important change: this new NRC edition includes requirements for ewes with singles, twins, *and triplets*, for all their periods of gestation and lactation. The previous edition contained no requirements for triplets. (Don't get me wrong — it's not that ewes raising triplets had no nutrient requirements; it's just that we didn't know enough about them to list any numbers). The new reference tables list gestation and lactation for ewes with "Three or more lambs." I'm comfortable with three, but the phrase "or more" may be a bit optimistic.

As a new feature for the dairy sheep industry, the lactation requirements now include sections for ewes *in milking parlors* — for all three stages of lactation.

The new NRC has also adapted to other recent changes in the sheep industry: the range of ewe weights is now much wider than in previous editions. The 1985 NRC listed mature ewes from 50–90 kg (110–198 lb). The new NRC lists ewes from 40–140 kg (88–308 lb). This is in recognition, no doubt, of the many flocks with smaller hair sheep breeds and also of those monsters I'm glad that I don't have to shear but undoubtedly someone does. Or perhaps this new upper weight category may reflect some Columbia-Bison crossbreeding trials that I have not yet heard about.

The NRC now includes a section on ram requirements, which is good. The ram production cycle is simpler than ewes: just maintenance and prebreeding. Either you are letting rams wander around the field doing nothing (*maintenance*), or you are pumping them up in preparation for their work period (*prebreeding*). Most rams also require extra nutrition during their post-breeding recovery period, but this new NRC lists no numbers for it. Which actually makes a lot of sense because rams exhibit so much individual variation in their level of emaciation by the end of the breeding season that providing a single table of nutrient requirements would suggest that one set of requirements fits all situations. This would overly simplify a complex situation and potentially result in inappropriate diets for these rams.

The requirements for young growing and finishing lambs are relatively straightforward. Table 15-2 contains different categories for early-maturing and late-maturing lambs, and for different ages and rates of gain. This table also lists separate requirements for ram lambs, because the NRC committee judged that ram lambs need slightly more energy for maintenance than ewe and wether lambs.

Now let's discuss replacement ewes and ewe lambs. The new NRC tables are extensive but somewhat confusing. I've studied these tables carefully, and here's my interpretation:

The NRC essentially describes three types of replacement ewe lambs: (1) farm flock ewes *first bred as ewe lambs*, (2) farm flock ewes *first bred as yearlings*, and (3) range flock ewes *first bred as yearlings*.

For farm flock ewes first bred as ewe lambs (to lamb at 12 months of age), their few months of growth after weaning are covered in Table 15-2 in the section on "Growing Lambs and Yearlings." Once they are exposed to rams at 35 weeks of age, their requirements are listed in Table 15-2 in the breeding and gestation sections under "Yearling Farm Ewes." After they lamb, their lactation requirements are listed in Table 15-1 in the lactation sections under "Yearling Farm Ewes." Then, after they wean their lambs but during the 19 weeks before they are put with a ram again, their requirements are listed in Table 15-2 in the "Maintenance Plus Growth" section under "Yearling Farm Ewes." Finally, when the ram is put back into the flock, their requirements are listed in Table 15-2 in the breeding and gestation sections for "Yearling Farm Ewes."

For farm flock ewes *not* bred as ewe lambs (so therefore they will first lamb at two years of age), their entire first year of postweaning growth is listed in Table 15-2 in the section on "Growing Lambs and Yearlings." Let's skip to the end of the second year: breeding and gestation are covered in Table 15-2 under the breeding and gestation sections for "Yearling Farm Ewes." The middle period — as old croppers after their first birthday but before they are bred — contains some months of growth, and therefore those requirements are listed in Table 15-2 under "Maintenance Plus Growth" and some months of simple maintenance while waiting for the ram. Those months are covered in Table 15-1 in the "Maintenance Only" section under "Yearling Farm Ewes."

For replacement range flock ewes, the new NRC simply doesn't believe that they will be bred as ewe lambs. The requirements for their entire first year after weaning are listed in the section for "Growing Lambs and Yearlings" in Table 15-2. The word "Yearlings" in the title suggests that these requirements also apply to these range ewe lambs until they are exposed to a ram at 18 months of age. After that, their requirements are listed in Table 15-2 in the breeding and gestation sections under "Yearling Range Ewes." Once they lamb at 24 months, their lactation requirements are listed in Table 15-1 in the lactation sections for

"Yearling Range Ewes" (even though they are actually older than two years of age).

Then again, if these range ewe lambs are *not* bred until they are older than 24 months, the requirements for their second year are listed in Table 15-2 under "Maintenance Plus Growth," or if they have reached their mature size, in Table 15-1 under "Maintenance Only." Once they are put with a ram in their third year, all their requirements for breeding and gestation are listed in Table 15-1 under "Mature Ewes."

Who's on first?

Well, I'll stop here. I can sense that your eyes are glazing over, some heads are beginning to nod off. No more pontifications, no more convoluted descriptions. Next month I'll change gears and discuss how the new NRC edition lists energy and protein requirements. Brace yourself — there are some new things here also.

First Published: April 2007

Author's Note: There are some very good things in this new NRC edition, but trying to trace the nutrient requirements of replacement ewe lambs and replacement yearling ewes in this new edition is rather challenging. We can only hope that enterprising computer programmers can incorporate these numbers into their ration-balancing software, so that the end-user doesn't have to search through the tables himself.

Energy & Protein, New

Let's continue with our description of the sheep nutrient requirements in the new Small Ruminant NRC book. This month, I'll concentrate on the reference values for energy and protein.

Although this NRC book is so important to our industry, the funny thing is that very few people will actually read it. Or even own it. At 362 pages retailing for $116, not many folks will give it as a birthday gift, and it probably won't be a best seller on Amazon. Another practical reason it is not widely read is that all its numbers are in the metric system. In the scientific community, the metric system rules, and all weights are in kg, g, mg, etc., and energy values are in calories, kcal, Mcal (kilo = thousand calories, M = mega = million calories) or even the cryptic megajoules per kilogram (MJ/kg). But on a working American sheep farm, how exactly would the shepherd feed 14 megajoules of alfalfa to a 70 kg ewe? So it is up to nutritionists in feed companies, universities, and government offices to read this book carefully and translate its values into pounds per day or percentages of dry matter or square light-years per gallon.

And speaking of energy, the NRC reference tables list the energy values of feeds and the energy requirements of sheep in terms of *ME — Metabolizable Energy*. What is this ME, how does it relate to something we know, like *TDN* (Total Digestible Nutrients), and why does the NRC use it?

Some basic nutrition here. Feed energy is expressed in calories per kg of dry matter, and the different energy terms reflect the successive energy losses that occur as feed and nutrients move from the feedbunk through the digestive tract to the cellular level. The initial energy value of a feed before it goes into an animal's mouth is called the *Gross Energy (GE)*. That's the feedbunk value. After a feed is consumed, the undigested portion is excreted in the manure. The GE minus the energy lost in the manure is called the *Digestible Energy (DE)* of the feed. This DE term is roughly equivalent to TDN, as I'll explain below. To continue the digestion process, some energy is then lost as gas (primarily from the rumen) as well as in the urine. DE minus the gas losses and urine losses is called *Metabolizable*

Energy (ME), which is the amount of energy that actually reaches the cells. But as cells use this ME, some energy is lost as heat. ME minus the heat loss is called *Net Energy (NE)*. It's this NE that is finally used by cells to do metabolic work such as maintenance, muscular activity, and the manufacture of tissues like muscle, bone, milk, and wool. To summarize this sequence: a feedstuff contains GE which loses fecal energy to become DE which loses gas and urine energy to become ME which loses heat energy to become NE.

Historically, American livestock books have described feedstuffs and nutrient requirements on the level of DE, as this term most closely relates to TDN and is also easily measurable (just by collecting manure). But much of the world currently uses ME, particularly producers in the UK, Australia, and New Zealand, so this NRC committee shifted its terminology from DE to ME. The most precise energy term, NE, is very hard to measure, so nutritionists generally feel more comfortable on the ME level. Although the new NRC does indeed list NE values for growing lambs, things about NE are very complex and questionable, and I would rather put off a discussion about it for another day.

So, how does ME relate to TDN? Although ME and TDN don't measure the same thing, we can still interconvert them by using a couple of rule of thumb assumptions along with a physical constant. One rule of thumb is that 1.0 lb of TDN equals approximately 2,000 kcal of DE. Another is that the sum of urine and gas losses equals 18% of DE, which means that ME is generally 82% of DE. The physical constant is that 1.0 kg equals 2.2 lb. Here's an example from the first row of Table 15-1: a 40 kg ewe at maintenance needs 0.41 kg of TDN per day. Multiplying 0.41 x 2.2 x 2,000 x 0.82 and then dividing by 1,000 (to convert kcal to Mcal) gives an energy requirement of 1.48 Mcal ME per day, which is exactly the number listed in the table.

This new NRC also refers to feeds with 8, 10, and 12 MJ of ME/kg. Whoa! What's this? Well, many countries describe their feeds, especially forages, in terms of MJ/kg, so this NRC committee has decided to include those values in the book, perhaps with the hope that we'll all someday travel to New Zealand. The trick of converting joules, calories, and TDN is in knowing the straight-ahead equivalency that 1.0 calorie equals 4.184 joules. Therefore, 1.0 Mcal ME equals 4.184 MJ ME. Here's an example: If a feed contains 66% TDN, I can start with this value (0.66 lb TDN per 1.0 lb of dry matter) and then multiply by all these factors to come up with a result of 10MJ/kg (I won't list all the arithmetic here — that is your homework assignment). Therefore, feeds containing 8, 10, and 12 MJ of ME/kg equate to 1.91, 2.39, and 2.87 Mcal ME/kg respectively, which equate to TDN values of 53.0%, 66.2%, and 79.5% respectively. Why do I even mention this? Because the fourth column of each requirement table lists dietary energy concentrations with these same weird numbers — 1.91, 2.39, or 2.87 Mcal ME/kg.

Let's switch to protein values, which may seem even more peculiar to most readers.

Most of us are comfortable with the concept of *Crude Protein*. We've used it forever as part of our daily vocabulary — buying and selling feeds, judging forages, and formulating rations. But this new NRC doesn't use Crude Protein, at least not directly. Instead, it bases its requirements on a number called *Metabolizable Protein (MP)*. MP is actually a better way of describing the complex reality of protein nutrition, but most folks are probably not familiar with it.

Metabolizable protein refers to the amount of protein actually absorbed across the gut wall. If we think about it, the concepts of crude protein and metabolizable protein are roughly analogous to the concepts of gross energy and digestible energy. The first concept of each pair represents the nutrient amount in the feed when it first enters the mouth; the second concept in each pair refers to the nutrient amount actually moving from the digestive system into the blood. (Disclaimer: for those of you who have a lot of knowledge in animal nutrition, you'll recognize that this is not a perfect match, but it's close.)

I'll be brief here: the crude protein in feed is composed of two parts: the portion that the rumen microbes can degrade and the portion which passes through the rumen relatively unscathed because the rumen bugs can't degrade it. The first portion is called *Degradable Intake Protein (DIP)* and the second portion is called *Undegradable Intake Protein (UIP)*. This second portion is also commonly known as *Bypass Protein*, although scientists tend to prefer the acronym UIP (possibly because it sounds more scientific). When degradable protein enters the rumen, things get rather complicated. Some DIP is lost to the animal because of microbial action, but most DIP is still available for absorption downstream in the lower tract. In contrast, the UIP is more efficient than the DIP because it "bypasses" the rumen and goes directly to the lower tract for absorption. MP is the sum of the amounts of DIP and UIP that are actually absorbed across the gut wall into the blood.

We need to know this background information because the nutrient requirement tables in the new Small Ruminant NRC contain *three* columns for protein requirements, not just one as in previous editions. These three columns differ in the percentage of UIP in total amount of dietary protein — 20%, 40%, or 60%. The higher the percentage of UIP, the more efficiently the crude protein is digested and the less total amount is needed in the feed. For example, a 70-kilogram mature ewe in late gestation carrying triplets needs 222 g/day of crude protein if her diet contains 20% UIP, 212 g if her diet contains 40% UIP, and only 203 g if her diet contains 60% UIP, while her MP requirement is simply 149 g.

To use these columns for ration-balancing, we have to estimate the amount of UIP in the diet. Values for UIP in feedstuffs can be found in this book's main feed ingredient table (Table 15-12) and also in other current textbooks and nutritional references. As a practical matter, however, when I look across all of the sheep categories, I notice that crude protein requirements seem to decrease by only approximately 10% as the percentage of dietary UIP rises from 20% to 60%.

These NRC reference tables also list requirements for MP and DIP. Although the DIP number is not very useful — it's just listed for reference — the MP number is the true basis for all the other protein requirements. The value for MP was calculated from underlying equations that add up the amounts of protein actually used for tissue growth and replacement. The NRC uses a simple formula to convert MP to CP, based on the percentage of UIP in the diet. It's clear that new ration-balancing software will use all these numbers to formulate rations, rather than just the traditional crude protein values of yesteryear.

It's also clear that I've run out of space. Next month, I'll cover the new NRC book from a very different perspective . . . of what coulda, shoulda, and woulda.

First Published: May 2007

Author's Note: We've come a long way from the days when someone would recommend a feedstuff or a diet because it really "put on the fat" or it gave a "hard finish." The energy and protein values in these tables were all derived from complex mathematical models, which in turn were derived from many and varied research trials, which in turn depended on, well, someone going out to the barn and feeding sheep. We can only hope the sheep appreciate all our work.

Tough Love

You see it in every showring — the judge lines up the animals and gives reasons. For each animal, he always starts with something good, and then he tactfully but systematically moves on to its shortcomings. Well, the new Small Ruminant NRC book was published this year, and I've spent the past three articles discussing its background along with a few main items. Now I'd like to cover some shortcomings — some obvious and some not so obvious. Not as a litany of complaints, but like a coach who applies tough love: he values the strengths of his athletes, but he also points out a couple of things that could make them better.

Why do this? Because this NRC is an incredibly important book. It's *the* no-nonsense nutritional reference volume for the small ruminant industries — sheep, goats, deer, and New World Camelids (alpacas, etc.). This book will be the bedrock of hundreds of publications, public and private. It will be relied upon by thousands of producers and students and their advisors — nutritionists, veterinarians, university professors, teachers, Extension agents, and government regulators. It will become the cornerstone of nutritional research for decades. So, as its influence grows, we should also maintain an honest appreciation for its imperfections.

Here's one of its most glaring problems: it contains *no* CD with the reference numbers in digital format. The most critical sections in this book are the 14 large reference tables covering 90 pages — listing the nutrient requirements for scores of animal categories and also the nutritional values of hundreds of feedstuffs. The sheep nutrient requirements alone take up 25 pages. If I wanted to use these numbers in a computer program for ration-balancing or in a reference document, must I type all these numbers in *again*? I am astonished. We live in a 21st century digital age, yet this book contains *no* CD-ROM. And I've checked the websites — as of this writing they don't offer any downloadable tables either. Other recent NRC reference books on beef cattle, dairy cattle, and horses all included digital supplements. Why not this one which contains even more tables than the others?

And speaking of tables, this NRC edition contains two large tables of feedstuff composition: Table 15-11 ("Common Feedstuffs") and Table 15-12 ("Other Novel Feedstuffs"). These tables are massive, each extending across 20 columns or more. Lots of data, lots of small print. But both tables are hard to read. I needed a ruler just to follow the numbers across the page. Some simple layout devices would have helped, like alternating shades of white and grey in different rows, or sequential line numbers, or entry numbers on both pages. These are not new concepts — other NRC books have routinely used them.

Another simple organizational device is the alphabet — as in *alphabetized order*. Table 15-11 indeed lists its entries in direct alphabetical order, so no problem there. But Table 15-12 is kind of a mess. Its title, "Composition of Pasture or Range Forage, Browse, and Other Novel Feedstuffs," suggests that it contains a potpourri of feedstuffs, and it does — from common forages like alfalfa silage to unusual trees like acacia to vegetables like artichokes. But Table 15-12 divides its entries into subgroups, like "Forages and Browse" and "Hays, Silages and Other Roughages" and "Vegetable Produce" and "Leafy Green Vegetables." Why divide feeds into subgroups? Why not simply list all the unusual feeds in alphabetical order so we don't have to guess the location? My guesses may not always coincide with the writing committee's design. For example, if I wanted information on tomatoes, in this table I might logically look under "Produce, Vegetables," but instead, the entry is listed under "High-energy Feeds" (and, I might add, wrongly spelled as "Tomatos"). Catagorizing feeds into subgroups is a throwback to some old nutrition textbooks which arbitrarily divided feedstuffs into various types of roughages and concentrates. Confusing indeed.

As for having *two* feedstuff tables — Hmmm. I appreciate the logic for separating common feedstuffs from novel ones — but I'm afraid these two tables don't do it well. Common feeds like alfalfa, turnips, and peas are listed in Table 15-12 but they shouldn't be. They belong in Table 15-11, where they should be listed only once. These are examples from the department of redundancy department.

Also, listing a feedstuff in two different tables just begs for discrepancies and errors. Every computer programmer will tell you to avoid entering the same piece of data twice. Here's an example: Table 15-11 lists the TDN and crude protein values for apple pomace (wet) as 68% and 6.0%, respectively, but Table 15-12 lists them as 74% and 5.6%. Which values should I use?

Now for some technical nutritional issues.

This NRC expresses energy requirements and nutritional values as both TDN and ME (Metabolizable Energy). It interconverts them by using the rule of thumb that 1.0 pound of TDN equals approximately 2,000 kcal of Digestible

Energy (DE) with a conversion factor of 82% from DE to ME. (I covered these concepts in greater detail in last month's article.)

In any case, this DE-to-ME conversion has a couple of problems. Firstly, the 82% conversion factor is not universal — it really only applies to fibrous feedstuffs. In contrast, we should use a conversion factor of 85% for grains and 79% for oil meal byproducts such as soybean meal. This non-uniformity was noted in the 1985 Sheep NRC as well as in the classic 1969 textbook *Animal Nutrition* by Maynard and Loosli. Also, the 82% conversion factor means that 18% of the feed energy is lost in gas and urine. In practice, I can reduce that loss by including an ionophore like Bovatec® or Rumensin® in the diet. These compounds alter the rumen fermentation to reduce gas production, which effectively raises the DE-to-ME conversion to above 82%. Also, animals with high intake levels or high amounts of fat in their diets would have lower gas losses because of reduced rumen fiber digestion, which in turn increases the DE-to-ME conversion rate. In any case, higher conversion rates translate to higher ME values relative to TDN. Therefore, these new NRC reference tables may slightly underestimate (by 3% or more) the ME values of certain feeds as well as the ME requirements of animals (or conversely overestimate the TDN values).

Then there is dry matter intake (DMI). This is a topic by itself, but here I'll be brief. Every table of nutrient requirements includes a column for DMI, but this column is not what you think it is. Remember that DMI is *not* a requirement, it's simply a nutrient "package." Those intake numbers are really just educated guesses. For each animal, the NRC estimates DMI from the animal's ME requirement and the energy value of a single ration. What about other ration formulations and other factors? The chapter on feed intake discusses dozens of factors that affect intake and also lists many complex equations. And most of those equations were derived from studies with sheep that were fed indoors. Also, when we try to apply these numbers to grazing animals . . . well, nutritionists know that the DMI variation on pasture is quite large. In essence, the NRC simply ignores this variation.

And finally, the NRC tables list DMI numbers to two decimal places (i.e., to the nearest 10 grams). Interesting, considering that this book contains a 13-page chapter describing the complex variables that make this number so hard to predict. Although *computers* may prefer many digits, do these decimal places really make the DMI number more meaningful or accurate? Two decimal places may give the appearance of credibility but it doesn't change the fact that the DMI number is still a guess.

There are alternatives. In his famous *Feeds And Feeding* books that for years were the backbone of every Animal Nutrition 101 course in the country, F.B. Morrison listed the nutritional requirements for livestock. But his DMI columns listed a *range* of values rather than a single number with two decimal places. Why? Because he knew animals and understood the practical issues of feeding them. He knew that DMI was not as predictable as we would like, but that people still needed to use DMI values in their practical situations.

This reminds me of a story about F.B Morrison when he was a professor at Cornell University. His assistants would help him compile nutritional information from all over the country by dutifully recording experiment results on index cards (remember index cards, those 3x5 pieces of stiff paper that pre-dated iPods?). F.B. Morrison would go through those index cards to extract the good information for his nutritional tables. How did he select the good information? In a very practical way that was based on shrewd observations and practical experience. He applied the "Morrison Criteria" to those index cards — which meant that if he found a card with numbers that did not make sense to him, he tossed it into the garbage can.

First Published: June 2007
Author's Note: Do these criticisms mean that I won't use this NRC reference book? No, of course not. But as I utilize this reference, I must also consider its shortcomings, which gives me a balanced view of the numbers and their limitations. The animals still have to eat, and we still have to work on the best ways of feeding them.

Diets, New and Old

When I recently used the new sheep nutritional requirements to balance some ewe rations, I noticed something disturbing. The new reference values seemed very different from the previous numbers I had been using for years. So I opened both reference books and investigated, calculated, estimated, and reiterated. And I found some things that were very interesting indeed.

Some background: the *National Research Council (NRC)* publishes the American reference tables for livestock nutrient requirements. The newest sheep requirements are listed in the 2007 NRC publication *Nutrient Requirements of Small Ruminants.* Although it covers sheep, goats, alpacas, and deer, this book contains an impressive 25 pages of tables for sheep requirements. The previous NRC publication on sheep — *Nutrient Requirements of Sheep, 6th Revised Edition, 1985* — was a slim volume that listed nutrient requirements on only seven pages. The copy on my bookshelf is dog-eared and heavily creased, like a well-worn pair of leather boots. Anyway, I'll refer to these two books as the *2007 NRC* and the *1985 NRC.*

Both editions explicitly list the nutrient requirements for ewes and lambs in various stages of production. The 2007 NRC contains more categories and wider ranges of body weights than the 1985 NRC, which is to be expected. The 2007 NRC also addresses some concepts in different ways, such as listing three different requirements for crude protein based on the proportion of *rumen bypass protein* in the diet. But I won't go off on these tangents. In this month's article, I'll just focus on comparing the old numbers with the new numbers.

Let's approach this comparison in a practical way. Instead of jumping around the chapters and comparing different formulas, kind of like kicking the tires of a new car, let's examine one typical feeding situation and compare the nutrient requirements listed in each edition. Specifically, let's choose a mature 154-pound ewe raising twins in early lactation. We'll assume that our ewe is in moderate body condition and not under any stress from weather or disease, so the numbers in these tables don't need any adjustments. By the way, 154 pounds is not a random weight I pulled out of a hat — 154 pounds equals 70 kg, and since all scientific tables are listed in metric values, each NRC book conveniently contains a line for a 70-kilogram ewe.

First, the old values. The 1985 NRC defines early lactation as the first *eight* weeks after lambing, and it assumes that during this period, a ewe raising twins will lose 0.13 lb (60 g) per day. To support milk production and prevent additional weight loss, the 1985 NRC lists this ewe's daily requirements for energy and protein as 4.0 lb TDN and 0.92 lb crude protein, respectively. For the main minerals and vitamins, our ewe needs 11.0 g calcium, 8.1 g phosphorus, 7,000 IU vitamin A, and 42 IU vitamin E. Based on these requirements, the 1985 NRC expects her daily dry matter intake (DMI) to be 6.2 lb, which is 4.0% of her body weight. A ration supplying her requirements at this level of intake would test at 64.5% TDN and 15.0% crude protein.

Now, the new values. As a literary device (what, in this article?) to make comparisons easier, I'll try to duplicate the exact sentence structure and wording from the previous paragraph. (Also, I've calculated my percentages from the original metric values, which may cause some rounding differences here.)

The 2007 NRC defines early lactation as the first *six* weeks after lambing, and it assumes that during this period, a ewe raising twins will lose 0.07 lb (31 g) per day. To support milk production and prevent additional weight loss, the 2007 NRC lists this ewe's daily requirement for energy as 2.88 lb TDN. Protein is a bit more complex because the 2007 NRC lists *three* different crude protein values determined by the percentage of rumen bypass protein in the diet (the 2007 NRC refers to bypass protein as *UIP — Undegradable Intake Protein* — but we will still call it bypass protein here). For diets containing bypass protein at 20%, 40%, and 60% of the total protein, the ewe's daily requirements for crude protein are 0.67 lb, 0.64 lb, and 0.61 lb, respectively, which equates to dietary protein levels of 15.5%, 14.7%, and 14.1%. For the main minerals and vitamins, our ewe needs 7.9 g calcium, 6.9 g phosphorus, 12,500 IU vitamin A, and 392 IU vitamin E. Based on these requirements, the 2007 NRC expects her daily DMI to be 4.4 lb, which is 2.8% of her body weight. A ration supplying her requirements at this level of intake would test at 65.4% TDN.

Wow, that's a lot of numbers. Now let's compare these two sets of requirements.

Interestingly, the main nutrient *percentages* are rather similar — both diets contain approximately 65% TDN and 14–15% crude protein. But these percentages are deceptive because they mask major differences in required *amounts* of nutrients. And when we do the calculations to balance diets, we usually try to supply the appropriate *amounts* to the animals, and the percentages are just byproducts of these calculations.

It is when we compare the required *amounts* of nutrients that we see stark differences. Compared to the 1985 NRC, the 2007 NRC requires 28% less TDN and approximately 30% less crude protein (for simplicity, I averaged the three 2007 crude protein values to 0.64 lb). Calcium and phosphorus requirements are 28% and 15% lower, respectively. And the 2007 NRC requires 79% more vitamin A and a whopping 833% more vitamin E. Actually, these last four values for minerals and vitamins are based on lots of new research, so I am confident that *these* differences are justified.

However, the 2007 NRC assumes our ewe will lose weight at only 0.07 lb/day (31 g), which is 48% less weight loss than indicated in the 1985 NRC.

Another item of real interest is the dry matter intake. The 1985 NRC predicts a DMI of 6.2 lb, while the 2007 NRC predicts only 4.4 lb — a 29% reduction in feed intake. These DMIs represent 4.0% and 2.83% of the ewe's body weight, respectively.

What gives? Well, let's see how these numbers were derived.

In the early 1980s, the scientists in the NRC Sheep Nutrition committee examined all the research to date, evaluated the pros and cons of these trials, looked at the outliers and averages and special circumstances, and poured through thousands of studies. The committee scientists were all experienced sheep men. Based on these studies and their own personal experiences, they tried to extrapolate these research results to practical situations and come up with reasonable numbers for each requirement.

In contrast, the Small Ruminants Nutrition Committee for the 2007 NRC pursued their task in a different way. Sure, the committee members looked at previous research and carefully evaluated the numbers. But these scientists *also* relied quite heavily on a complex computer model — the *Cornell Net Carbohydrate and Protein System for Sheep (CNCPS-S)*. A computer model is nothing but an extensive set of interlocking equations, with coefficients and variables, and behind each equation lies a set of assumptions. The CNCPS-S model is complex and flexible; it's designed to predict responses when the inputs are known. For the beef cattle and dairy industries, similar models have provided some reasonable results.

Each number in the 2007 NRC reference tables was built step by step from component parts, with this computer model contributing input at various steps along the way. Particularly for energy, protein, DMI, and weight loss.

But I wonder . . . the discrepancies are larger than I expected. Compared to the 1985 NRC numbers, the 2007 NRC reduces energy and protein requirements by 30%. Do these smaller amounts of energy and protein really provide enough nutrition to support milk for two lambs while simultaneously reducing ewe weight loss — especially in the face of lower feed consumption? My experience tells me that ewes will consume 4% or more of their body weight during lactation. This new model predicts less than 3%. If the new model is accurate, the wide differences in recommendations between the two NRC books imply that those scientists in 1985 were rather inaccurate in *their* judgments.

In the 22 years between editions, have modern ewes changed so much? Or have we learned so much more? Or have we just altered our approach to estimating nutritional requirements, so that these editions reflect changes in the technique rather than actual requirements? Or don't we know?

We need some good research. Animal Science Departments and USDA Experiment Stations throughout the country should conduct dozens of bread-and-butter studies in which sheep are fed diets formulated for the requirements listed in these two NRC editions. Head-to-head comparisons, so to speak. The results should reveal which assumptions are correct and which requirements are accurate.

Otherwise, without such comparisons, we could be in a situation best described in a quote from the 1995 movie *Apollo 13*, "Houston, we have a problem."

First Published: January 2009
Author's Note: Am I condemning the new 2007 NRC? No, not at all. But I am asking a serious question. Does the model in the 2007 NRC really work? There is a lot of opportunity for graduate-level research in that question.

Applying the Principles

It's a dangerous world out there. Calories and protein come at you from all directions. Especially the calories. Nearly every magazine and newspaper carries articles about nutrition; obesity is a headline event. Shoppers everywhere pause in supermarket aisles to read package labels. Whoa! I'm a livestock nutritionist, not a food purveyor. I *can* make sense out of all this, but only in the framework of livestock nutrition. So let's apply the principles of livestock nutrition to some common supermarket products and see where it leads . . .

I'll start with the vocabulary of energy and protein. In livestock nutrition, especially for ruminants, we often report the energy value of feeds in terms of TDN (*Total Digestible Nutrients*). We can convert TDN to Calories by using the rule of thumb that 1.0 pound of TDN equals 2,000 Calories of *Digestible Energy*, which is the feed energy actually absorbed across the gut wall.

The other main nutrient is protein, which is listed on every animal feed tag as *crude protein*. It's well-known in the livestock world that analytical labs don't measure the amount of *true protein* in a feed, they only measure the amount of *nitrogen*. They then multiply this number by 6.25 to get the value for crude protein. That's why it's called *crude* — to remind us of what it's not.

And what about dry matter (DM)? This is an important concept. In livestock nutrition we always describe a feed's nutritional value on a *dry matter basis*, which means "without the water" because pure water contains no nutrients. The *as fed* nutritional values are always lower than the DM values because the *as fed* values are diluted by the water in the feed. The conversion, however, is simple arithmetic. We correct for the amount of water by dividing the *as fed* values by the dry matter percentage as expressed as a coefficient. This is easier to demonstrate than to explain: If a hay contains 90% dry matter, we convert the *as fed* values to a DM basis by dividing the *as fed* values by 0.90. For example, if the crude protein level of that hay was listed at 14% on an *as fed* basis, then the hay's crude protein value on a *DM basis* would be 15.6% (= 14 ÷ 0.90).

To round out our nutritional principles, we must remember some physical constants: 1 g of pure carbohydrates contains 4 Calories. Same with 1 g of protein (from a nutritional point of view). On the other hand, 1 g of fat contains 9 Calories. And 1 pound of anything equals 454 grams (and therefore 1 ounce equals 28 g).

For practice, let's first apply these concepts to three typical livestock feeds: alfalfa hay, shelled corn, and high-quality grass pasture. I just happen to have these feed reports handy, so here are their respective nutritional values on a DM basis: 62%, 88%, and 70% TDN, and 18.0%, 10.0%, and 24.0% crude protein. What does one pound of each contain, on a DM basis? (We'll convert TDN to Calories and report protein in grams. You can do the math at home.) One pound of this alfalfa hay contains 1,240 Calories and 82 g protein; one pound of this corn contains 1,760 Calories and 45 g protein; and one pound of this grass pasture contains 1,400 Calories and 109 g protein.

Now let's apply these concepts to human feeds . . . er, foods. (And while we're at it, perhaps we should now designate the dinner table as a food trough.) The package label of nearly every supermarket food item contains a well-marked section called "*Nutrition Facts.*" How convenient — kind of like a feed tag. Anyway, let's examine some of these as if we were livestock nutritionists. We'll express the results in terms of the composition of one pound of dry matter.

Let's start with one of most common foodstuffs: table sugar (typically sold in a 5-pound bag). Its *Nutrition Facts* lists a serving size of 4 g with the following nutrient amounts per serving: 0 g fat, 0 g protein, and 4 g carbohydrates (specifically, it identifies these carbohydrates as 4 g of sugars — not exactly a surprise). Let's assume that this sugar is 100% dry matter, so we'll make no correction for DM. Therefore, 1 lb of sugar contains 1,816 Calories (= 4 Calories per gram x 454 g) and 0 g protein. More energy than corn but definitely not a good protein supplement. And of course, no energy from fats, trans or otherwise.

Let's look at another common food — I've just pulled a loaf of whole wheat bread from my shelf. Its *Nutrition Facts* lists a serving size of 43 g (= one slice. Okay, I'll bite. Just to check, I weighed a slice on my postal scale, and it is indeed 43 g). This serving contains 21 g carbohydrates, 1.5 g fat, and 4 g protein. Hey, this only adds up to 26.5 g, not 43 g! What happened? Water. Each slice of this bread contains 16.5 g water. Meaning arithmetically that this bread contains 61.6% dry matter (= 26.5 ÷ 43 as a percentage). And conversely, 38.4% water. Maybe that's why bread companies can boast that their breads are soft and moist?

In any case, these 26.5 grams of dry matter contain all the carbohydrates, fat, and protein in the serving. Let's first analyze the protein. Four grams of protein in 26.5 g dry matter equates to 69 g protein in 1.0 lb dry matter (454 g), which means that this bread contains 15.1% protein — not as good as alfalfa hay or good pasture but better than shelled corn. Energy is a little more complex because this foodstuff contains carbohydrates *and* fat. Our 26.5 g dry matter contains 21 g carbohydrates and 1.5 g of fat. Using basic ratio analysis techniques, we extrapolate these numbers to 1 lb (454 g) of dry matter — and find that this pound contains 360 g carbohydrates and 26 g fat. Applying our Calorie conversions of 4, 4, and 9 respectively, we calculate that this pound contains 1,950 Calories (= 4 x 360 + 4 x 69 + 9 x 26). Why more Calories than straight sugar? Because of the fat. This bread contains 5.7% fat (= 1.5 ÷ 26.5). And at 9 Calories per g, a little fat goes a long way.

Here's an example of just how long this way can be: another item pulled from my shelf is a bag of corn chips. (I keep these chips on that shelf only for research purposes, of course.) Its *Nutrition Facts* lists a serving size of 28 g (= 1 ounce, which is about nine chips). These 9 chips contain 19 g carbohydrates, 2 g protein, and 6 g fat, which total 27 g dry matter — which means that these chips are at least 96.4% DM. I say "at least" because the chips also contain some salt and other minerals, but for the purpose of this article we'll just focus on the main nutritional components, and we'll also round the numbers for convenience.

The 2 g of protein in this 27 g of dry matter equates to 34 g protein in 1.0 lb DM, which is a 7.4% protein feedstuff. If you're running a feedlot, that's even less protein than shelled corn. Yes, I know — not many folks consume corn chips for its protein. They like the fat, of course. That 6 g of fat doesn't sound like much, but hang on. The 19 g carbohydrates and 6 g fat in each serving translate to 319 g carbohydrates and 101 g fat in 1.0 lb. Again, using our 4-4-9 conversions, we calculate that the carbohydrates, protein, and fat contribute 1276, 136, and 909 Calories respectively, which totals to a whopping 2,321 Calories in 1.0 lb dry matter. Speaking of feedlot rations . . .

And 39% of these 2,321 Calories come from fat. That's because this foodstuff contains 22.2% fat (= 6 ÷ 27), and fat contains 225% of the energy of carbohydrates or protein.

It's a good thing that the "serving size" listed on the bag is only 9 chips, and that most people carefully limit their intake to 9 chips.

As livestock nutritionists, we may notice a couple of things that are slightly awry. Firstly, supermarket labels list the term *protein*, not crude protein, even though their laboratories test for nitrogen in foods in exactly the same way as agricultural laboratories test for nitrogen in livestock feeds. Of course, most human foods don't contain urea or other forms of nonprotein nitrogen, so this discrepancy may be only marginally significant.

But here is something more consequential: Supermarket labels list "Calories" without defining *which type of Calories*. Are they Calories of *Digestible Energy*, as in livestock feeds? Actually, no. Supermarket labels list Calories of *Gross Energy*, which is the number of Calories going into the mouth — with no correction for digestibility or absorption. Which means that on a supermarket label, *all* carbohydrates have the same caloric value of 4 Calories per gram, whether they are digestible or not. Which means that table sugar, which is a highly digestible soluble carbohydrate, has the same caloric value as pure cotton thread, which is the indigestible fiber carbohydrate that we call *cellulose*. No wonder that the results of many human weight-reduction diets are rather variable.

Food for thought. Meanwhile, pass me that spool of cotton thread. Mmmm. At least it doesn't have any calories from fat.

First Published: July 2007

Author's Note: Applying the principles of livestock nutrition to supermarket labels can be frightening. Concepts like dry matter, caloric density, digestibility, fiber, etc. — all are very basic in Livestock Nutrition 101, but all are totally ignored in the supermarket labels that everyone reads. Fascinating.

SECTION 5

Feed Tests, Feed Reports, Feed Tags

Black Box Forage Analysis

Forage analysis isn't what it used to be. A new technology called NIR is revolutionizing how we analyze feeds and forages. It's fast, portable, and very impressive. But NIR isn't all ease and glory — sometimes you should avoid it altogether . . .

NIR? National Institute of Redundancy? Nutritionally Improper Research? Never Interested Rams?

Actually, NIR stands for *Near-Infrared Reflectance Spectroscopy* (sometimes also written as "NIRs"). NIR equipment analyzes forages by using sophisticated electronic circuitry rather than old-fashioned chemical reactions.

The guts of the NIR system is a *spectrophotometer*. This machine emits light beams of carefully selected wavelengths and then precisely measures the amounts reflected from the sample. You may remember "ROY G BIV" — that old mnemonic we used in high school to memorize the colors of the visible spectrum (hint: R = red, V = violet). Wavelengths get longer as you go from V to R, and wavelengths longer than red are called *infrared*. Infrared wavelengths nearest to the red region are called, logically, *near-infrared*. Every feedstuff reflects near-infrared light in different amounts. The NIR spectrophotometer measures this reflected light.

Measurement, however, is only half the story. The other half is translating these measurements into useful numbers. Raw data coming off a spectrophotometer looks incomprehensible — a jagged line of blips across the page — kind of like the graph of the federal deficit. The NIR computer transforms this data by using some very sophisticated mathematics and prediction equations. Then it prints out protein, fiber, and other nutrient values in numbers that we can understand. (Frankly, I have a suspicion that the computer simply takes the square root of the rancher's birth date and multiplies it by the second derivative of his driver's license number.)

So why is NIR touted so highly in the nutritional world? Firstly, NIR is quick. After a technician prepares a feed sample by drying and grinding, he can obtain a complete NIR nutritional analysis in 5 minutes, counting printing time. That's fast. Traditional wet-chemistry methods require glassware, boiling liquids, and forced-air ovens. Although many laboratories have automated their wet-chemistry procedures, wet chemistry is still much slower than NIR.

Secondly, NIR is portable. The entire apparatus consists of a microwave oven for drying samples, a small grinder, the spectrophotometer (which is a little larger than a microwave oven, but not much), a computer, and a printer. The whole kit and caboodle can easily be trundled into a van and hauled to educational meetings, hay auctions, and even farms and ranches. Try that with a room full of glassware.

Surprisingly, NIR was originally used back in the 1960's by commercial grain traders to measure oil, moisture, and protein levels in wheat and other commodities. In 1976, animal nutritionists published the first papers demonstrating that NIR can also be used to predict forage values. Extension Services in Wisconsin, Minnesota, and Pennsylvania soon purchased mobile NIR units. In Wisconsin, the NIR van virtually revitalized hay auctions. Hay dealers learned that they could have their loads analyzed on the spot and make more money by selling hay that was objectively measured for nutritional quality. Many universities soon obtained NIR equipment, and now private commercial laboratories are purchasing NIR equipment.

Exciting? Yes — but here's the rub: *NIR does not truly analyze anything real.* Unlike traditional chemical methods, NIR measurements *do not* represent exact chemical bonds or quantities of atoms. NIR measurements are simply reflectance spectra of selected wavelengths. The trick is to establish a predictive relationship between these spectra and real world nutritional values.

We routinely do this through a process called *calibration*. This is the development of the unique prediction equations that the NIR computer uses to interpret the raw data. Every forage lab must run through its own calibration process when it first sets up its NIR system. A lab derives its calibration equations by statistically comparing benchmark feed samples to their NIR spectra. In other words, in each lab, someone tediously analyzes certain standard feedstuffs using old-fashioned wet chemistry methods. Then they analyze duplicate samples of these same feedstuffs using NIR equipment. Then they correlate these numbers and construct prediction equations that relate the NIR spectra to their wet-chemistry values. Once a lab is confident that its correlations are stable and repeatable, it uses the NIR equipment to analyze unknown feed samples, such as the hay from your ranch.

But there is a caveat: NIR accuracy is limited to the types of samples used for calibration. This is a very important principle. For example, if a laboratory meticulously derives calibration equations for alfalfa hay, we can be fairly confident that its NIR equipment can accurately predict the nutritional value of alfalfa hay. Labs routinely derive calibration equations for many types of forages and other feedstuffs, but they cannot calibrate for everything. Prediction equations derived from alfalfa hay cannot be reliably applied to ryegrass-clover hay or fresh oat pasture or even alfalfa silage. Unfortunately, if these samples are analyzed by NIR, the computer will use its best-guess equation and will still generate nutritional values for them. The problem is that these nutritional values may be wrong.

Everyone has unusual forages. "Unusual" in this sense means non-routine from the laboratory's viewpoint. A forage lab in Oregon may be well-prepared for analyzing ryegrass-clover hays, but it probably won't have calibration equations for sudangrass hay or culled lima beans or puna chicory. Similarly, a Minnesota lab may routinely and accurately analyze typical grass hays from the Midwest, but I wouldn't trust its NIR values for bermudagrass grown in southern Georgia. "Unusual" doesn't only mean different species — unusual means anything not in the lab's NIR calibration curves, even things like drought-stressed alfalfa or very highly fertilized grass.

There is a solution, however. If you have an unusual feedstuff, *just request wet chemistry methods.* For example, if you send a sample of sudangrass hay to a lab that doesn't have calibration equations for sudangrass, don't automatically rely on the lab's NIR procedures. Instead, ask the lab to use wet chemical methods on your sample. Most labs have this option. Wet chemistry may be old-fashioned, and it may cost a bit more than NIR, but for non-standard samples, the results are more dependable.

First Published: October 1994

Author's Note: Even with the ongoing improvements in technology, this entire process still relies on good calibration. Being aware of the limitations of NIR is important and practical. Blindly relying on the numbers in a lab report may result in an expensive mistake if those numbers don't reflect the true nutritional value of the feed.

Anomalies

Anomalies. Aberrations. Abnormalities. Unexpected deviations. Webster's dictionary defines *anomaly* as "*the state or fact of being out of place, out of true, or out of a normal or expected position.*" Well . . . are there any anomalies in agriculture? Are there ever! Here are some to think about . . .

A good place to begin is the anomaly between forage tests and soil tests.

Look at any forage test report. The values for phosphorus (P) and potassium (K) are always listed as percentages of the dry matter, as in 0.16% P and 2.3% K. Microminerals are expressed in ppm rather than percentages because their levels are so much lower (for example, 9 ppm copper) — that's why they have the appellation *micro*. Now look at a soil test report. The P and K values are listed as ppm, as in 18 ppm P and 200 ppm K. The anomaly is *not* the different amounts of P and K — we expect that soils will contain lower levels of P and K than feedstuffs — but that these values have *different meanings*. The P and K values in a forage test do *not* mean the same thing as the P and K values in a soil test.

Think about what happens when a laboratory analyzes a feedstuff. The procedure always requires some boiling or cooking or dissolving the sample in various solutions, followed by isolating the fraction of the sample that contains the mineral, and then using clever techniques such as light emission or enzymatic reactions to quantify the mineral in that fraction. The procedure uses the entire feedstuff in the analysis. The resulting number represents the *total amount* of that mineral in the feedstuff, every atom of it.

A soil test, on the other hand, takes a completely different approach. A soil test is not interested in every atom in the soil; it's only interested in the atoms *that are available to plants*. And herein lies the anomaly.

Consider a typical soil sample: it consists of particles of clay, silt, sand, and also some organic matter. Most of the mineral atoms are firmly locked within the crystal compounds of these particles. From a plant's perspective, atoms locked inside soil particles might as well be on Jupiter for all the good they do. What matters to plants — and to farmers and graziers — are the mineral atoms that dissociate from those soil particles and leach into the soil solution, because plant roots can *only* absorb things that are in solution. For example, soil particles may contain a lot of P, yet typically less than 1% of this P is available in the soil solution.

Soil testing procedures, therefore, attempt to mimic the dark conditions of soil by extracting minerals into solution and then measuring those extractable levels. The resulting number — for example 18 ppm P or 200 ppm K — represents the amount of mobile atoms in solution that plants can absorb.

Let's compare these two approaches. A feedstuff value of 0.16% P can be also expressed as 1600 ppm, which is actually a rather low level of P in a feed. Yet a soil test P of 200 ppm would be a high level for soil. We just need to remember that these numbers represent different things.

Speaking of soil and fertility brings me to another important anomaly: fertilizer labels. Look at the numbers on the side of a fertilizer bag — for example, 16-16-16-6. Don't these numbers represent the percentages of nitrogen, phosphorus, potassium, and sulfur? Not really. The first and last numbers *do* represent the percentages of nitrogen and sulfur, respectively. But the middle two numbers represent percentages of *phosphorus oxide* (P_2O_5, phosphate) and *potassium oxide* (K_2O, potash), not P and K. Some arithmetic shows that phosphorus atoms only constitute 44% of P_2O_5, and potassium atoms only constitute 83% of K_2O. Therefore, expressing the *true* percentages of the four minerals in 16-16-16-6 would give us a label that would read 16% N, 7% P, 13% K, and 6% S.

But you'll never see *those* numbers on a fertilizer bag.

Why not? If feed tests and soil tests report the values of elemental minerals, why not fertilizer labels? Another anomaly. This one is attributable to history. Over a hundred years ago, before our high-tech computer gadgetry, scientists used rather unsophisticated methods for measuring the mineral content in fertilizers. They ignited the product and weighed the resulting oxides at the bottom of the crucible — in other words they burned their sample — and then, instead of converting numbers back to the elemental forms, they reported the results as oxides — P_2O_5 and K_2O.

Historically, once things get started, they tend to have a momentum of their own. As the reporting of fertilizer values as oxides became more widespread, people began to rely on those numbers more often. Oxide values soon became the accepted language in the fertilizer industry. Field advisors from the Extension service and fertilizer companies published their recommendations in terms of those values. State regulatory officials then wrote their labeling laws to *require* oxide values on fertilizer bags. Thus, over time, the use of P_2O_5 and K_2O became accepted throughout the industry and fossilized into legal code.

But do plants absorb or use P and K as oxides? *No.* Do nutritionists express animal nutrient requirements as oxides? *No.* Do fertilizer companies want to change the convention of listing P and K as oxides? Again, *no*. And why should they? If P and K were listed as percentages like N and S, their label numbers on a fertilizer bag would be *lower* than in the current system. Which is not exactly a popular proposal with folks in the business of selling fertilizer. And so it goes.

And finally there is the anomaly of cats and vitamin B_{12}.

Cats, of course, need vitamin B_{12} as a required nutrient. Nothing weird here. Ruminants such as sheep and cattle also need B_{12}, but for them there is a difference. Remember that in ruminants, healthy rumen microbes manufacture B-vitamins which are subsequently moved downstream through the gastrointestinal tract and absorbed into the blood from the animal's small intestine. Ruminants, in effect, possess a self-contained fermentation system that supplies their essential B-vitamins. Yes, you are reading this correctly — under normal conditions, sheep and cattle do not require supplemental B-vitamins like biotin, thiamine, or niacin.

Vitamin B_{12}, however, is the slight oddball exception for ruminants, because each B_{12} molecule contains one atom of cobalt. Rumen microbes can create this B_{12} molecule only if they have access to cobalt atoms. Actually, cobalt has no other metabolic role in animals except as a component of the B_{12} molecule. Therefore, while we don't feed intact B_{12} molecules to ruminants, we do routinely include cobalt in their trace mineral mixtures to supply cobalt to the rumen. This strategy keeps the rumen microbes happy and allows them to supply B_{12} to their host. Look at a typical mineral feed tag for cattle or sheep, and you will usually see the ingredient *cobalt carbonate* or a similar cobalt compound.

Which brings us to cats. Most of us recognize that cats are not ruminants. (Coughing up hairballs is assuredly not the same as chewing cud.) In fact, I suspect that most cats would be insulted if they were mistaken for cows.

Nutritionally, since cats don't have rumens or rumen microbes, they must obtain fully pre-formed B_{12} molecules from their diet. Therefore, we routinely add B_{12} to their food by including ingredients like "Vitamin B-12 Supplement" or "cyanocobalamin" (a technical name for B_{12}).

But . . . you should go to the pet food section of your local supermarket and read the labels on cans of cat food. All of them contain some form of Vitamin B_{12}, of course, *but some cat foods also include ingredients like cobalt carbonate or cobalt chloride.* Why? Well, that's a good question. It's another anomaly. Feeding cobalt to cats does no good because only rumen microbes can convert cobalt into vitamin B_{12}. But maybe those feed companies have secret metallic interests in cats that we don't know about.

As Webster succinctly put it, an anomaly is *"the state or fact of being out of place . . ."* Perhaps we need a bumper sticker that says, *"So many anomalies, so little time."*

First Published: January 2003
Author's Note: These anomalies are fun to think about. I suppose that one could spend a professional lifetime ranting against such inconsistencies, but frankly, I've learned that the world is not exactly a perfectly logical place. I enjoy discovering these little imperfections in my profession.

What Do I Look For?

Some folks, I suppose, look at a forage test and see only a list of numbers. That's kind of like looking at a map of Wyoming and seeing only a diagram of roads. When I look at a Wyoming map, I see snow-covered mountains, open range, and the metallic blue of a wide Wyoming sky.

Forage test reports are like short stories. Each is a narrative of old friends, surprising plot twists, and sometimes sly interactions between characters. So let's turn our attention to forage tests, and I'll share with you some of the things I see.

All forage test reports — and I'll use this term to mean all feed tests — contain two columns of numbers. One column is labeled "As Sampled" or "As Fed" or "Wet Basis" or something like that, and the other is almost always labeled "Dry Matter." Which one should we look at?

I'll start with a practical situation. Let's say that we have two feeds: Feed A contains 58% TDN and 13.4% crude protein; and Feed B contains 16% TDN and 3.3% crude protein. Which feed is better? Of course this is a trick question — otherwise why would I ask it? And of course, anyone looking at these raw numbers would choose Feed A.

Now the punch line: Feed A is early-grass hay. Feed B is cow's milk. *Now* which feed would you choose? Of course, milk is the better feedstuff, but these numbers don't show it. Why? Because these numbers are "as fed" values — right out of the barn, as they are fed, *which includes the water in those feeds.* Milk contains 87.6% water, and this water dilutes the other nutritional values, making them smaller. In this case *a lot* smaller. The hay in our situation contains only 11.0% moisture. Since these feeds contain different amounts of water, comparing their nutritional values on an "as fed" basis is like comparing apples and oranges. Water contains no nutritional energy or protein, so we must eliminate it from the report. All laboratories, therefore, adjust for this water by dividing the "as fed" number by the dry matter coefficient. Then they list the corrected results in the "dry matter" column of the report.

Let's make that adjustment to our two feedstuffs. Our results now show that, on a dry matter basis, Feed A (the grass hay) contains 65% TDN and 15.1% protein, and Feed B (the cow's milk) contains 129% TDN and 26.6% crude protein. *Now* these numbers reflect the *real* nutritional values of the feeds. So, in a forage test report, which column should we look at? I *always* look at the dry matter column.

When I first receive a forage report, I quickly scan the entire page, trying to get a full picture of the feed. I look at the general values for crude protein, TDN, horse TDN (HTDN), calcium, phosphorus, and dry matter. Then I ask: Are these numbers close to what I expect for this type of forage? I'm searching for weird numbers, unexpected values. I recently received a forage test that reported 37% protein for a grass pasture. 37% ?!! Wow — *that* number certainly caught my eye. I asked the lab to retest the sample. They did, and the results came back unchanged. Only then did I have real confidence in the accuracy of that number.

After scanning the entire report, I then scrutinize the energy value. I like to use TDN, but many reports also list values for digestible energy, metabolizable energy, and/or a whole raft of numbers for net energy (maintenance, growth, lactation), and also a Relative Feed Value. That's okay. Remember that labs don't actually test for energy; labs only test for *fiber* (generally ADF = *Acid Detergent Fiber*) and then plug that fiber value into an equation to *derive* the energy value of the forage. For example, labs use a formula that looks something like "TDN = A *x* (ADF) + B", where the values of A and B come from reference tables for different types of feeds. That's why it's so important to label a forage sample accurately. Lab technicians use that label to determine which values of A and B to apply.

A few years ago in Oregon, a rancher sampled some perennial ryegrass hay for analysis. On the submission form, he wrote the words "p. hay" as the label for his perennial ryegrass hay. When I received a copy of the report, I noticed some unexpected values for energy and protein, so I called the lab. The technician pulled up the form and we found the problem — the lab thought that the sample was *pea* hay and used reference values for legumes rather than grasses. H-e-l-l-o? This is a clear example of the need to label clearly. After our conversation, the lab technician ran the numbers through the computer again, this time with the proper category, and the report's nutritional values were more appropriate.

Labeling samples properly is even more critical if the lab analyzes the sample with *NIRs* (*Near Infrared Reflectance Spectroscopy*) rather than old-fashioned wet chemistry. NIRs is, literally, an electronic black box technology that uses infrared waves and sophisticated mathematics to deduce nutritional values. With NIRs, *all* of the numbers, including the values for both types of fiber (NDF and ADF), are derived from reference tables. Samples with inaccurate labels can show wildly fascinating results in the final report — good for laughs but not much else.

Then I look at crude protein, and if it is listed, I look at the supplemental number for something called *adjusted crude protein* or its inverse, *ADF-N*, which is nitrogen bound up in the ADF and not available to the animal. These supplementary numbers can indicate heat damage — which can occur when wet hay is stacked in a barn. Some of the forage protein can cook into a gooey, indigestible substance like caramel. This shows up in the forage report as ADF-N. Some labs list ADF-N directly, while other labs translate this nitrogen into crude protein in the usual way (by multiplying by 6.25) and list *unavailable crude protein*, which is then subtracted from the total crude protein to arrive at a value for adjusted crude protein. Either way, a high number for ADF-N or a low number for adjusted crude protein, relative to the total crude protein, tells me that the forage had suffered some protein losses due to heating.

In practice, nearly all forages have a tiny amount of nitrogen naturally linked to fiber, which reduces the adjusted crude protein value slightly. This is no big deal, and I generally ignore it. But if the adjusted crude protein is *significantly* lower than the value for total crude protein, then I balance rations with the adjusted value, because *that* represents the biologically usable amount of protein. How much is significantly lower? That depends on the level of crude protein, but anything more than one or two percentage units is significant.

And then I look at . . . uh, oh . . . we're out of time and space.

Well, come back next month, and we'll look at the minerals listed in forage tests. Fascinating stuff, minerals. You might say that they are, uh, the salt of the earth . . .

First Published: March 2001

Author's Note: This is only a start, of course. Laboratories often list many different energy and protein nutritional values in their reports. You can always ask a nutritionist for the fascinating details about the meaning of each intriguing value. But in practice, if you look in the correct column and focus on a few critical numbers, you've already come a long way in understanding the true nutritional value of that forage or feed.

Looking at Minerals

Let's continue to look down the dry matter column of a forage test report. We've already examined the array of values for energy and protein, so what's next? The minerals. Lots of interesting numbers. Here are some things that I consider when I look at those mineral levels.

I first look at the levels of calcium and phosphorus, and I am particularly interested in two things: (1) the absolute levels of these minerals, and (2) the ratio between them. Grasses typically contain 0.25–0.60% calcium (remember that everything in this month's article is expressed on a *dry matter basis*). Legumes tend to have higher calcium levels than grasses, often 1.4% of the dry matter or higher. For example, if I see a forage calcium level of 1.2%, I would guess that this forage contained a high percentage of legume. Phosphorus levels for both grasses and legumes tend to be in the range of 0.15–0.50%. Low phosphorus levels tell me that the forage was quite mature. High phosphorus levels, on the other hand, may be due to high levels of phosphorus fertility in the soil. But if I see values outside of these ranges, I look very carefully at them. For example, I would ask if something unusual occurred in that field, like a fertilizer spill, or if the forage lab did the correct assay.

For the ratio between calcium and phosphorus (Ca-P ratio) in the *total diet*, I like to see at least 1.3:1 for most situations and, ideally, 2:1 or slightly higher for young, growing animals. Once the phosphorus requirements are met, having a calcium level of approximately twice as high as phosphorus helps insure that male animals don't suffer from a syndrome called *urinary calculi*, in which insoluble crystals containing these two minerals form in the urethra and block urination. Older animals generally don't need such high Ca-P ratios, but under some conditions, ratios lower than 1:1 may cause calcium deficiency problems in adults, particularly the *milk fever* syndrome in mature ewes or dairy cows. Since many feed companies add calcium and/or phosphorus to their supplements and mineral mixtures, knowing the forage levels of calcium and phosphorus helps me understand the total diet and guides my decisions about the need to supplement them.

After evaluating the calcium and phosphorus numbers, I look at the level of magnesium and an associated mineral, potassium. Low magnesium levels contribute to the spectacular neurological problem of *magnesium tetany*, which is also called *grass staggers* or *winter tetany*, among other names. Whatever we call it, symptoms occur when blood magnesium levels drop below a trigger threshold, which causes the animal to go into seizures. We usually see this problem in the early spring when forage is lush and young. The rule of thumb about magnesium tetany is rather simple: *low risk* for forage magnesium levels above 0.20%; *moderate risk* for levels between 0.15–0.19%; and *high risk* for levels below 0.15%. Except that . . . (drum roll, please) . . . high potassium levels in a forage can reduce magnesium absorption from the intestinal tract. How high? Another rule of thumb: forage potassium levels above 3.0% can cause problems, particularly if forage magnesium levels are low or marginal. Compared to legumes, grasses are particularly greedy about potassium — they will absorb extra potassium from high-potassium soils, even above their own requirements for growth. I've seen grass test higher than 4.0% potassium, and in the early spring before the soil really warms up, the primary forage growth is grass rather than legumes. So knowing the levels of magnesium *and* potassium helps me evaluate the metabolic risks of magnesium tetany and the option of adding extra magnesium to the mineral mix during the risk period.

Some folks ignore the level of sodium, but I don't. Most trace mineral mixtures contain white salt (sodium chloride) which also usually acts as the main palatability factor that drives animals to consume the mixture. High sodium levels in forages make me a little nervous, because those high levels may satisfy an animal's desire for salt and thus reduce its intake of the free-choice TM mixture. Forages usually contain less than 0.20% sodium, but I've seen levels higher than 0.40%, especially in forages grown near the ocean or on saline ground, and also in byproduct feedstuffs and other supplemental feeds.

And then there is copper. This is a big bugaboo, especially among sheep producers who rightly worry about chronic copper toxicity. But I'm not only just interested in copper, I'm also interested in three other minerals that affect copper absorption: molybdenum, sulfur, and possibly iron.

Sheep, cattle, and goats all have a nutritional requirement for copper, at approximately 8–11 ppm in their total diet when the dietary molybdenum level is low. But sheep are particularly sensitive to chronic copper toxicity, and even slightly higher copper levels over a long period could cause problems. I'm generally happy to see 8–11 ppm copper in a forage test, and this seems to be a common range in forages. But what about forages grown in old orchards, where farmers periodically sprayed trees with Bordeaux mixture (copper sulfate + hydrated lime + water), or in fields where hog manure or chicken litter was applied as fertilizer? What about feeds composed of copper-containing ingredients or feeds mixed incorrectly? I always want to know about elevated levels of copper, and copper values greater than 15–18 ppm are red flags that I look at very carefully.

But copper absorption is profoundly influenced by molybdenum and sulfur, and to some extent, iron. High dietary levels of these minerals will reduce copper absorption across the gut wall, and if they are high enough, may even cause a copper deficiency. Forage molybdenum levels can range from less than 1 ppm up through 3 or 4 ppm or higher. Sulfur levels are generally 0.10–0.30%. I would consider sulfur above 0.35% to be high.

In an ideal world, I would like to see a ratio of copper to molybdenum (in the total diet) of between 6:1 and 10:1. Higher ratios may increase the risk of copper *toxicity*, especially if the sulfur levels are also low, and lower ratios suggest the possibility of copper *deficiency*, especially if sulfur levels are high. Iron can also tie up copper, so I am wary of high iron levels, say above 400 ppm. On the other hand, high iron can also be due to soil contamination of the sample, which is not necessarily a nutritional issue, so I take these iron levels with a grain of salt.

Forages contain other required minerals, of course, such as zinc and manganese. I always scan these numbers for uncommonly high or low values, looking for obvious problems. Selenium, iodine, and cobalt values would also be useful, but most laboratories don't test for them.

Someday, however, I would like to see laboratories provide information about some unusual minerals, like radium and uranium. Because if a forage contained high levels of these minerals, I could feed that forage knowing that I could always find my animals at night.

First Published: April 2001

Author's Note: This is the second article in a series about reading forage test reports. The first focused on the numbers for energy and protein. This one concentrates on the macro and micro minerals. I would note, however, that three important required minerals — selenium, iodine, and cobalt — are not listed in most commercial laboratory reports. Which means that, in practice, we either make assumptions about their levels or spend extra money getting specialized analyses.

Feed Tag Bestsellers

No college English professor ever required students to read feed tags as their assigned readings. Well . . . okay . . . maybe feed tags are not as exciting as *Macbeth* or *Paradise Lost,* certainly not exciting reading to an English professor. Feed tags are short, pithy, and full of numbers. They have poor sentence structure and no plot. But in the practical world of livestock nutrition, feed tags *do* tell us what we are feeding our animals and also what we are *not* feeding. If we're spending thousands of dollars each year on feed, maybe we should know these details. So let's make believe that we are still in school, and the professor has assigned a reading list of feed tags.

By law, all commercial feeds must have feed tags, and all feed tags must have the same basic design. Ironically, there is no national feed tag law that governs this system. In the U.S., each state controls its own feed tag regulations. But fortunately, all state laws are similar because they all follow the same pattern — the *Model Bill* designed by the national organization *Association of American Feed Control Officials* (AAFCO). Since 1910, the AAFCO has worked with all sections of the feed industry and the FDA to create its *Official Publication,* which lists feed tag specifications in mind-boggling detail. Each state, in turn, writes its own feed tag laws based on these specifications. This consistency is a good thing too, because major differences in feed tag requirements would wreak havoc with the interstate movement of commercial feeds.

In any case, all feed tags contain the same sections. The top front side shows the names of the manufacturer and the feed. Incidentally, the AAFCO rules for naming feeds are designed to prevent problems of misrepresentation. The rules state that one or two numbers standing alone in a feed name must signify the protein value of that feed, even if that name doesn't contain the word "protein." For example, if a feed is named something like "Super 16 Complete Feed", we can safely assume that it contains 16% crude protein. But if those numbers do not stand alone, as in "Super Complete Feed 16B", those numbers may signify the company's catalog code number for the feed rather than a protein value.

If the feed contains any drug, the word "Medicated" *must* appear clearly and alone under these top items. Beneath the word *Medicated*, the feed tag must then list the purpose of the feed (i.e., "complete feed for growing wolves"). The feed tag must then list the name, dosage, and purpose of that drug.

Some notes about drugs: the word *Medicated* in a feed tag means that the feed contains one or more of the FDA-approved drugs for that livestock species. Conversely, if a feed tag does not show the word *Medicated*, you can be sure that the feed does *not* contain any of these drugs. Details about the drug(s) are listed in the *Active Ingredients* section of the feed tag. The words in this section are precise and legal — the exact phrasing approved by the FDA for this compound, including its official name, dosage, purpose, claims, withdrawal time, and warnings. Companies must print the phrases approved by the FDA without change. Remember however, that the official name of a drug may be different than its popular name. For example, the ionophore *Bovatec*® is often included in sheep feeds, but the *Active Ingredients* section will list this compound as *Lasalocid Sodium*, because that's the official FDA-approved chemical name.

Only drugs approved by the FDA are listed in the Active Ingredient section. Surprisingly, some well-known feed additives do not fall into this category, such as probiotics, enzymes, and various microbial products. Although these items are often added to commercial feeds, they are not officially classified as drugs. Years ago, the FDA decided that these items fell under the category "Generally Recognized as Safe" (GRAS), and therefore did not need official approval. This means that the manufacturers of these GRAS additives, unlike the manufacturers of other drugs, are not required to prove to the FDA the efficacy of their claims or dosages.

The next two sections of the feed tag are the *Guaranteed Analysis* and *List of Ingredients*. The final parts of a feed tag, which may be printed on the back side, include the feeding directions, any warnings or caution statements (i.e., "Do not feed to lactating tigers"), the name and address of the manufacturer, and a quantity statement (i.e., "Net Wt 50 lb").

The *Guaranteed Analysis* section of a feed tag makes particularly good reading, because this section lists the actual nutritional value of the feed — the levels of protein, fat, minerals, vitamins, etc. These are the numbers we can take to the bank, or in some cases, to court if there is a problem with the feed or if the analyzed nutritional value doesn't match the feed tag value. Again, the AAFCO *Official Publication* gives very specific requirements about this section — the values that must be listed for each type of feed, and even their sequence within each list. For example, a complete broiler feed would have a different array of nutritional values than a mineral mix for sheep.

These guarantees, however, are not *equal* signs — they are stated as *minimums* and *maximums*. So if a feed lists a minimum percentage of 16% for crude protein, the feed may legally contain more than 16%, although economics probably makes this unlikely. But there would be real problems if this feed's protein level analyzed at only 12%.

Most feed tags list the levels of various minerals. Selenium is certainly a critical item, although some types of feeds may not be required to list it. For free-choice trace mineral mixtures in the U.S., the current legal limits for selenium are 90 ppm for sheep and 120 ppm for beef cattle. A little history here: feed tags now list selenium in ppm (parts per million). It wasn't always so. Older tags listed selenium as a percentage, and therefore we'd see numbers like 0.0050%, which was hard to compare against the ppm values of nutrient requirements and legal limits. But there is a simple trick — and you don't even need a computer to do this. To convert percentages to ppm, *just move the decimal four places to the right*. Therefore, 0.0050% selenium in a mineral mix converts to 50 ppm, which makes it easy for us to judge that this value is rather low for many regions in the country.

And then there is the issue of *fiber*. The AAFCO *Official Publication* requires that fiber levels be expressed as *Crude Fiber*, which for ruminants is so inaccurate that the number is useless except perhaps to lawyers. More than 25 years ago, nutritionists learned that the newer fiber values — NDF and ADF — were far more accurate and valid than crude fiber for predicting the true nutritional characteristics of a feed. Today nearly all certified laboratories use NDF and ADF to derive the energy levels listed on their test reports. Frankly, the only reason that crude fiber doesn't cause more problems to livestock producers is because many commercial feeds are primarily composed of grains which contain little fiber, and therefore the effect of this inaccuracy is relatively low. For ruminants, crude fiber is an antiquated value that really belongs in a museum, not a feed tag. And yet labeling laws still require it on the feed tag. Hmmm . . . I wonder what my old English professor would think?

First Published: January 2001

Author's Note: I've only touched on the broadest points of the Model Bill here. In the Official Publication of the AAFCO, the section on feed tag regulations is more than 90 pages long. Lots of details, some major, some arcane — but they are all important in situations where they apply. We should remember that feed tag rules apply not only to livestock and poultry feeds, but also to feeds for rabbits, dogs, and cats, as well as other specialty feeds.

The Small Print

I enjoy giving presentations about feed tags. I display a few sample feed tags on a large screen, and I explain the significance of each of the terms, section by section, including the list of ingredients. During nearly every presentation, someone in the audience inevitably exclaims, "Oh, *that's* what that stuff is!" Let's talk about some of the stuff you find in the *Ingredients* section.

Nearly every feed tag has an *Ingredients* section which simply lists the feedstuffs in that feed. This section is different from the *Guaranteed Analysis* section, which lists the feed's nutritional quality. The *Ingredients* section contains, one by one, the legal names of each ingredient the company puts into the bag. Well, more or less. More about this below. But one thing to remember — there is *no* national feed tag law in the United States. Each state develops and publishes its own feed tag regulations. But to our relief, the state laws are not a chaotic jumble of 50 widely divergent rules. All states base their regulations on the prototype *Model Bill* that is printed in the official reference book by the Association of American Feed Control Officials (AAFCO). There are some minor differences between states, of course, but not many.

Let's discuss the commonalities. The *Ingredients* section can only list the *official* names of feedstuffs as defined in the AAFCO publication. For example, this publication gives a precise description of the common feedstuff "cottonseed meal," including the criterion that it must contain at least 36% protein. But portions of the AAFCO publication can make for some fascinating reading. In addition to the common feedstuffs, it contains detailed descriptions of feed ingredients not exactly familiar to most of us, like reed-sedge peat, parboiled rice bran, and dried shellfish digest.

More commonalities: ingredients can only be listed without reference to any grade or brand name. For example, corn must be listed as corn, not #2 shelled corn by XYZ milling company. Another rule — familiar to everyone who has used word-processing software — is that all names must appear in the same typesize. This prevents feed companies from masking certain ingredients by listing some of them in large 12-point type and others in unreadable 4-point type.

Here's a universal principal for feed tags: *all ingredients in a feed must be listed*. Period. If something is not listed in the *Ingredients* section (excluding drugs which are listed elsewhere on the feed tag), *it has not been included in the feed*. Except for, uh, three exceptions.

First exception: if a feed contains only one ingredient, the feed tag does not have to list it. In their administrative wisdom, regulators figured that if the large letters at the top of the bag spelled "corn," farmers could reasonably assume that the bag contains corn rather than pickled olives.

Second exception: those things that come along for the ride because they are part of other ingredients. Specifically, things like minerals and vitamins. Manufacturers must always list mineral and vitamin components *when they add them as separate items*. For example, if a feedmill adds dicalcium phosphate or sodium selenate to a feed, you'll see these names on the feed tag. But we also know that most feedstuffs contain some minerals and vitamins. These components will not be listed separately in an ingredient list. So let's apply deductive logic: since corn contains phosphorus, a feed containing corn must also contain phosphorus, even though the feed manufacture did not add a *separate* phosphorus mineral to that mixture and no phosphorus mineral ingredient is explicitly listed on the feed tag.

This may seem obvious and even trivial, but it's not. The biggest cause for concern here is copper. It is well-known that sheep can be especially sensitive to chronic copper toxicity, and therefore sheep producers often look for feeds with little or no copper. They carefully scan the *Ingredients* sections for any words containing the terms *copper* or *cupric* or *cuprous* (the Latin names for copper). If these copper terms are not listed, a producer can only be assured that the feed contains no *added* copper. The operative word here is *added*, because *the feed can still contain copper* — the amount depends on the copper levels in the other ingredients. Some molasses products, for example, may contain more than 60 ppm copper, which would certainly elevate total copper levels if the ration contained a lot of that molasses. The concept of "not added" does *not* mean "zero."

The third exception applies to those peculiar ingredients called *Collective Terms*. These are not individual ingredients; they are umbrella terms that represent *multiple* ingredients. The AAFCO publication lists seven collective terms (some states don't permit all seven, but all states permit at least some of them): animal protein products, forage products, grain products, plant protein products, processed grain by-products, roughage products, and molasses products.

Think of collective terms as *families* of ingredients. Feed laws precisely define these collective terms and explicitly list the ingredients represented by each one. For example, *processed grain byproducts* can represent 39 different feedstuffs, including dried brewers grains, corn grits, oat groats, peanut skins, rice polishings, wheat bran, and wheat red dog (now *that's* an intriguing name). If the collective term *processed grain byproducts* appears on a label, then that ration contains one or more of those 39 ingredients.

Collective terms are not a sinister plot to trick farmers; they are actually important tools in the feed industry. Feed manufacturers rely on collective terms to give them flexibility in formulating feeds. All feeds must meet certain constraints: nutritional specifications, costs, physical characteristics, palatability, etc. Feed companies are very aware that profits depend on their reputation. A ration made from inexpensive ingredients will not help a reputation if the animals refuse to eat it. So feed companies and their nutritionists engage in a delicate balancing act — they juggle ingredients, prices, availability, nutritional value, physical attributes, and taste — and they use collective terms on the feed tag to give them some freedom to mix and match ingredients in response to market fluctuations and opportunities. This is no small thing. Without collective terms, feed labeling would be an economic and administrative nightmare, and feed prices would be much higher.

But notice that there is something *not* listed in the *Ingredients* section: the actual percentage of each ingredient. Feed tags of commercial livestock feeds are "closed" — which means that the actual proportions of the ingredients are *not* listed on the label. The opposite situation — "open" feed tags — is routine for private custom mixes, but not for commercial feeds. Closed tags are also standard, of course, for human feeds (er, foods). The next time you are in a supermarket, look at a package of hotdogs or breakfast bars and try to find the exact percentage of each ingredient.

But all is not lost. We can still ascertain the *relative amounts* of the feed ingredients because the ingredients are usually listed in descending order of inclusion, by weight. The first ingredient is in the highest proportion, the second ingredient is in the second highest proportion, etc. Ironically, most state laws do not require this descending order for livestock feeds, but they do for pet feeds. And also for human foods. So if "water" is the first ingredient in a list, you may be buying a product with a whole lot of water.

By the way, in case you were wondering, *reed-sedge peat* is indeed a real feedstuff. The AAFCO Official Publication describes it as "a natural, partially decomposed plant material formed from a mixture of reeds, sedges, grasses and some hypnum mosses occurring in wetlands and containing one third to two thirds peat fibers." It's actually used in animal feeds as a carrier of liquid products or as a filler to lower the energy value of a diet. But it can only be included at levels less than 5% of a ration. Whew.

The *Ingredients* section — small print or large print, it's well worth reading.

First Published: February 2001
Author's Note: Knowing how to read feed tags can save you money and grief. I routinely recommend to my clients that they retain every year's feed tags in an envelope marked with the year label. That way, we have a chance of going back after the fact and sleuthing through a nutritional problem. We've solved a lot of nutritional mysteries that way.

SECTION 6

Sheep Production Cycle Nutrition

Decisions

Of all the decisions we make in a sheep operation, our most important decision involves the numbers 15-15-6-8-8. Huh? These numbers aren't the snap count of a football quarterback. These numbers apply to ewes, and they overwhelmingly influence profit. Still don't know? Here's a hint: they add up to 52. Getting warm?

Okay, here's the scoop. These numbers describe the annual production cycle of a ewe, in weeks, and they give us a framework for balancing rations. The NRC reference tables categorize a ewe's nutritional requirements by these different periods.

Now for details: A ewe's year can be divided into her different stages of production, with each stage lasting a fixed number of weeks. These stages of production are maintenance (15 weeks), gestation (21 weeks), and lactation (16 weeks). A ewe in *Maintenance* is a dry ewe: she is neither pregnant nor lactating. She has very low nutritional requirements, because all she does all day is walk around, eat grass, grow wool, and perhaps jump a few fences. The *Gestation* period, however, is more complex. Since more than 60% of fetal growth occurs during the last trimester of pregnancy, especially during the last six weeks, nutritionists divide gestation into *Early Gestation* (15 weeks) and *Late Gestation* (6 weeks). Similarly, nutritionists divide *Lactation* into *Early Lactation* (8 weeks), which includes the period of high milk production, and *Late Lactation* (8 weeks), when milk production declines rapidly. Therefore, the phrase *15-15-6-8-8* describes the periods Maintenance, Early Gestation, Late Gestation, Early Lactation, and Late Lactation, respectively. Each period has its own set of nutritional requirements.

This framework gives us a blueprint for feeding our ewes, and like any blueprint, it can help us make our sheep operations more efficient.

For example, let's consider weaning, especially early weaning. We know that the nutrient requirements of ewes in late lactation are higher than ewes in maintenance. Look at the NRC requirements for a 154-pound ewe raising twins (154 lb = 70 kg). In late lactation she requires 3.6 lb TDN and 0.73 lb protein each day but during maintenance she requires only 1.5 lb TDN and 0.25 lb protein. If we assume that her daily DM intake (based on the NRC requirements) is 5.5 lb during late lactation and 2.6 lb during maintenance, her total intake during these 23 weeks is 581 lb (= 308 + 273).

But what happens if we wean the lambs at 8 weeks of age? Early weaning eliminates the late lactation period and converts it into maintenance, therefore increasing the maintenance period from 15 to 23 weeks. Our ewe's total intake for a 23-week maintenance period would be only 419 lb, a savings of 162 lb DM, 118 lb TDN, and 27 lb protein. (Homework assignment: do these calculations yourself. Use both sides of the paper if necessary.) We would still need to feed the weaned lambs, of course, but that feed would be used more efficiently because it would go directly into the growing lambs. This is a business decision, and now we have a good handle on its nutritional implications as well as its direct feed costs and savings.

Before we continue, let's examine two crucial points. First, *what is the period of highest nutritional requirements*? I ask this question at many workshops. No one ever says Maintenance, but some folks occasionally say Late Gestation. No, it's not. Actually, the period of highest requirements is Early Lactation. Look at the reference tables for our 154-pound ewe with twins. Her daily requirements are 2.8 lb TDN and 0.47 lb protein for Late Gestation, and 4.0 lb TDN and 0.92 lb protein for Early Lactation. It's Early Lactation, no contest.

The second point is that for most sheep operations, *feed expenses represent more than 70% of the total budget.* I use "feed" in its global business sense, not just the out-of-pocket expenses for grains and mineral, but *all the expenses* necessary to provide nutrition to the sheep — the land you purchase or rent, taxes on that land, fencing to confine sheep in pastures, equipment to harvest hay and feed it, barns to store hay, labor to feed it out and haul away bedding and manure, etc. All these items add up. You can see that most of the time and resources in a sheep operation really go towards providing feed for the animals.

Which brings us back to the main topic: What is the most important decision in a sheep operation? The answer is "When is lambing?"

This answer involves our 15-15-6-8-8 framework of nutrient periods. *Because when we choose a flock's lambing date, we automatically fix all its nutritional periods and therefore most of the costs of the sheep operation.* Gestation comes just before lambing, lactation comes immediately after lambing, and maintenance occurs 8–16 weeks after lambing, depending on the weaning date.

These connections seem obvious, so what's my point?

My point is that many sheep operations in the U.S. and Canada choose to lamb during the winter. Actually, most sheep publications over the past twenty years have routinely recommended winter lambing — so that producers could market their commercial lambs before the traditional price slide during the summer and fall.

But consider this: when you choose winter lambing, *you automatically assign the ewe's periods of highest nutritional requirements to the months when you must rely on stored feeds — hay, silage, and grain.* The implications of this decision are profound. If you rely on stored feeds, then you have to store those feeds somewhere on the farm. You need to raise those feeds or buy them or both. But to meet the high nutritional requirements of gestation and lactation, you must provide good nutrition, which means that your winter feedstuffs should always be of high quality. If your home-grown forages are not high quality, then you are forced to obtain supplemental grain or high-quality forages from off the farm. How easy is it to make good-quality hay? What are the costs of obtaining good quality feeds? What are the costs of *not* obtaining them? How much risk is in this system?

Winter lambing also requires a lambing barn or shed, unless you are living in a mild-winter region like the Deep South or the coast of the Pacific Northwest. A lambing barn means lots of labor, as well as higher risks for some of our favorite diseases like pneumonia and scours. A lambing barn also severely limits the size of your flock, unless you can obtain skilled labor for lambing, which in turn makes you dependent on keeping that labor.

So from a business perspective, we must ask ourselves this question: is winter lambing *really* worth it? Is the potential for higher prices worth (1) the guaranteed higher costs of feeding pregnant and lactating sheep during the winter, and (2) the inherent risks and limitations of this type of operation? And conversely, how much faith can we put in the dual assumptions of high spring prices and low summer prices? After looking over the records of monthly lamb prices during the past few years, I wouldn't want to bet the farm on such a predictable price curve, at least not in North America.

Spring lambing, on the other hand, reserves the cold months for maintenance and early gestation — periods of low nutritional demands. When ewes lamb in the spring, their early lactation period coincides with the explosive growth of high-quality spring forage, and their breeding occurs during mid-autumn, which is generally their period of highest fertility, and when heat stress is not an issue.

Am I saying spring lambing is for everyone? No, not at all. Some producers may choose winter lambing for very good reasons — such as meeting a specialized lamb market, or using available labor, or taking advantage of a climate that allows high-quality forage growth in the winter. An excellent choice based on sound business judgments.

Yogi Berra once said, "When you come to a fork in the road, take it."

So each year, when you are thinking of when to put the ram in with the ewes, you are coming to a fork in the road. You can take the early route to winter lambing or the late route to spring lambing. It's your decision.

First Published: February 2002

Author's Note: This article was based on the numbers listed in the 1985 Sheep NRC reference tables. The new 2007 Small Ruminant NRC has divided the ewe's yearly cycle into slightly different categories, especially with three periods during lactation (early-mid-late lactation at 6 weeks, 6 weeks, and *X* weeks afterwards, respectively, where X ends when the lambs are weaned). But gestation still comes before lambing, and lactation still comes after lambing, and the critical decision of lambing date versus nutritional needs still has not changed.

One more thing. This article was the 100th monthly *"From The Feed Trough..."* column that I wrote for *The Shepherd*. Whew! Lots of things to ruminate on. Thank you, dear reader, for bearing with my ideas and sense of humor all these years, and thank you Guy and Pat Flora and Kathy and Ken Kark for your continued service to our industry and your unwavering support of my writing.

Differences Among the Lambings

People often talk about accelerated lambing and out-of-season lambing as if they are the same thing. They are not. Although ewes can lamb out-of-season without being accelerated, the converse is not true — accelerated ewes *must always* lamb out-of-season at least once during their cycles. Managing ewes in an accelerated program is more complex than managing ewes to lamb out-of-season once each year. Not only must shepherds focus on issues about fertility and estrous cycles, but they also must change their nutritional strategies. Accelerated lambing ewes require a nutritional management that is radically different from that of any other sheep.

The phrase *out-of-season lambing* simply means that a ewe gives birth during a non-normal lambing period. In the northern hemisphere, ewes usually do not lamb between June and November. Because a ewe's gestation period is approximately 147 days, out-of-season lambing means that conception must occur between January and June. If a ewe lambs out-of-season — but only once each year — she follows a nutritional schedule that is simply offset by a few months from the standard nutritional schedule. Since the main management problem of out-of-season lambing is convincing ewes to cycle during their off-season, scientists have tried hard to manipulate their hormonal patterns — by using breeding tricks, light treatments, hormone injections, hormones as feed additives, etc. On the other hand, the nutrition for these ewes is relatively easy. Just locate enough feed of the appropriate quality, fulfill the nutrient requirements for the standard nutritional cycle, and make money. No big deal.

Wait a minute . . . *standard nutritional cycle*? What do I mean? Well, sheep nutritionists — being a little bit different than say, human nutritionists — view the ewe's nutritional year in terms of physiology rather than months or seasons. Ergo, they've outlined a standard ewe nutritional cycle which reflects its five physiological states: (1) maintenance, (2) early gestation, (3) late gestation, (4) early lactation, and (5) late lactation. For a ewe lambing once each year, these periods are typically 15, 15, 6, 8, and 8 weeks in length, respectively. (If you do the math, these periods only add up to 364 days, but what's an extra day between friends?) The actual lambing event, of course, occurs between late gestation and early lactation, at least on every farm that I've ever visited, but it has no real bearing on this nutritional calendar.

Let's focus our discussion on two periods of this calendar — *maintenance* and *late lactation* — because accelerated lambing affects these two periods the most.

Maintenance is simply the 15-week period when a ewe is neither pregnant nor lactating. She spends her time growing wool, chewing cud, mowing pastures, and otherwise using up feed. Nutritionally, not very demanding. It's really a period of nutritional overhead, of marking time.

Late lactation is the final 56 days of a 112-day milking period. Milk production in most ewes drops off quickly after 60–90 days, especially relative to the needs of fast-growing lambs. Therefore, late lactation, particularly after 90 days, is really a period of camaraderie between the ewe and her lambs, not a time of nutritional stress on the ewe. It is also a period in which ewes can regain the weight lost during early lactation.

Once-per-year lambing, therefore, *whenever* it occurs during the year, means that a ewe's relatively non-productive periods comprise 44% of her annual cycle (= 105 days maintenance + 56 days late lactation as a percentage of 365 days). Those 161 days represent a rather high nutritional overhead — particularly for big ewes that don't produce much wool.

Accelerated lambing, on the other hand, is a whole new ball game. Accelerated lambing means that a ewe must lamb more frequently than once-per-year, *which mathematically requires that her standard nutritional year be compressed.* Accelerated systems force sheep producers to solve reproductive *and* nutritional problems.

Consider the Cornell STAR system, in which ewes can lamb five times in three years. The reproduction hurdles are obvious: ewes must conceive every 7.2 months, which automatically means that some lambings will be out-of-season. Each cycle of the STAR system includes pregnancy, which cannot be compressed, and early lactation, which remains unchanged at 56–60 days. But in the STAR system, the lambs are weaned at 60 days, and after a short 12-day recovery period, the rams are then reintroduced. And what happens to the periods of maintenance and late lactation? They vanish.

The STAR system thus eliminates that 44% nutritional overhead. But the flip-side is that the STAR system presents a serious nutritional challenge. While the standard once-per-year lambing gives ewes ample opportunity to recover body condition during the periods of late lactation and maintenance, accelerated systems give no such grace period. In an accelerated lambing system, the thin ewes at the end of lactation may not come into heat during the next accelerated period and they will be hard-pressed to recover their body condition. Accelerated lambing is a tough game — there is no slack time for nutritional recovery or easy living.

Thus, accelerated lambing systems require a continuous supply of high quality feed — energy, protein, minerals, vitamins, the works. I'm not implying that accelerated ewes should be fed until they are fat. Not at all. Just that these ewes must never lose their working condition. Accelerated lambing means that a rancher must plan his feed supplies at least six months in advance — plotting long-term moves with the same kind of strategy used for playing a game of chess — using efficient feeding facilities and good nutritional knowledge.

And pastures? Once-per-year lambing systems provide all sorts of opportunities for managers to juggle things like pasture yields, forage species compositions, and ewe body conditions, regardless of when the ewes lamb. Accelerated lambing systems, however, severely limit these options. Because accelerated ewes cannot be allowed to lose body condition, they require good feed all the time. Accelerated ewes must graze only high quality forages. No overstocking, no heavy weed infestations, no empty mineral feeders. Grazing accelerated ewes is like grazing lactating dairy cows — a manager must pull these ewes from fields much sooner than most people would normally consider for sheep, like when pastures may still contain *more than 1,500 pounds* of residual dry matter per acre. Managing these pastures properly may require the addition of other types of livestock who have lower nutritional requirements than these ewes, such as feeder cattle or replacement dairy heifers.

Finally, in trying to feed accelerated ewes, I would carefully question the accuracy of published nutritional requirements. The standard NRC reference tables were derived from once-per-year lambing schedules, which include periods of maintenance. These reference values are quite adequate for simple out-of-season lambing, but how do they apply to accelerated ewes? What *are* the true nutritional requirements for accelerated ewes? *Certainly they are not lower than the current NRC recommendations*, but if they are higher, then how much higher? Hmmm . . . we're getting into relatively uncharted territory here.

In practice, if I had an accelerated lambing system, I would feed my ewes better than ever before. I'd obtain information from other accelerated flocks rather than flocks that only lambed out-of-season. And I would definitely maintain an open mind.

First Published: November 1996

Author's Note: The new 2007 Small Ruminant NRC contains some important differences in sheep nutrient recommendations from the older 1985 NRC, but the basic principles of the nutrition for accelerated lambing ewes has not changed. In fact, the 2007 NRC still does not contain any nutritional reference values for accelerated lambing ewes.

The Numbers Game

You're rushing to catch a plane at O'Hare airport in Chicago. The check-in line at the ticket counter is long, but you're not worried because there are three ticket agents on duty.

But then you look carefully at that front counter. The first counter position is moving folks through smoothly, so no worries there. But at the second counter position, a person is trying to buy a complex roundtrip ticket to Auckland, with stopovers in St. Louis, Phoenix, Acapulco, Tahiti, and Rarotonga. (Are there any livestock in Tahiti?) The ticket agent is trying to locate Rarotonga on a map. And at the third counter position, an angry passenger is fussing over two pieces of missing baggage. He just got off a direct flight from Denver to Chicago, but one suitcase has ended up in Istanbul. The other suitcase is apparently missing, and the computer has no record of it.

Now you begin to worry.

Yes, physically, there are three ticket agents behind the counter, *but in reality, only one agent is actually servicing the line.* The other two agents are effectively occupied and out of the picture during the time you need them. So . . . what are your *real* chances of actually catching your flight on time?

Let's relate this situation to a sheep operation at breeding time. Breeding often occurs during the summer for an early winter lambing. Remember that the actual length of time that ewes come into heat (*estrus*) is only 30 hours or so. If for any reason a ewe is not serviced during her estrus, she has to wait for the next cycle — *at least 16 days*. It's kind of like catching a flight — if you're late, you're too late.

Let's say that you've just put out three rams with 100 ewes. Are there *really* enough rams for those ewes? Or is it possible that you have a situation like the lines at the airport, with only one ram providing actual service?

Most sheep producers know the old rule of thumb about having three rams per 100 ewes — so that two can fight while one does the important work. Yet, is this always enough? Think back to previous years. Was each lambing season tightly confined to a three-week period? Why not?

It's really a matter of arithmetic and timing. And also geography. The basic premise is that a ram cannot be in two places at the same time, particularly if those places are separated by 200 acres or by high hills. For a successful short breeding period, there must be enough fertile, active, willing rams to find and cover every ewe during the precise few hours that she is in heat. At any given moment, regardless of the numbers of ewes and rams in the paddock, the micro-ratio of fertile rams to a particular willing ewe must be 1:1. If for any reason this arithmetic is not met, that ewe will wait another 16 days.

What can prevent effective service? Let's concentrate on the rams. Firstly, there are the classic physiological problems, like epididymitis, poor semen quality, foot problems, poor body condition, and injuries. We can usually detect these with annual breeding soundness exams.

Secondly, there is heat stress. While rams under heat stress pant a lot, they still may work. Unfortunately *4–6 weeks later* their sperm may not work. High temperatures can damage sperm cells in the early stages of development, and those sperm cells won't reach the end of the tunnel, so to speak, for a few weeks. (Heat stress won't make much of a difference if all the ewes are already bred by the time the poor quality sperm becomes available).

Then there is ram behavior. Recent research at the USDA Sheep Station in Idaho has clearly shown that ram sexual interest is a variable. This means that, regardless of looks, conformation, or "style," a ram may not be a ram may not be a ram. In other words, some rams have wildly aggressive libidos — the "just-let-me-at-'em!" attitude that we all like to see in rams. Some rams are more mellow about their ewes — they are more selective or shy or affected by poor weather conditions ("sorry honey, it's raining too hard"). And some rams are so laid back that they fall into the low-libido category.

And what about pecking order? What happens if one ram dominates the others because of age, size, or sheer orneriness, and if that dominant ram doesn't have a high libido?

And then there is sexual preference. Most rams prefer ewes; some rams aren't very interested in sex; and some rams prefer other rams exclusively. This is new research. We don't know how common these behaviors are, but we do know that they exist.

Now let's look closely at our sample 100-ewe flock. How many rams are actually on service at any one time? Of our three rams, it turns out that one ram is infertile but nonetheless it aggressively dominates the pecking order. The second ram is quite willing and fertile, but it often fights with the first ram. The third ram doesn't have much libido and only breeds on alternate Wednesdays. Therefore, when a ewe comes into heat, only the second ram will service it, and he may be otherwise occupied with fighting the first ram. Or if the paddock is too large or hilly, he may not even find that ewe in time. And thus the breeding season stretches on and on . . .

And how does all this affect nutrition? Simple — an extended breeding season wrecks havoc with nutritional management. In general, four months after you turn in the rams, during late gestation, you begin feeding grain. Or should you? This is a dilemma: if you start feeding grain to the early-lambing ewes, the late-lambing ewes will initially receive extra feed for too long and will probably become too fat by lambing time. A long lambing period also means that you'll have lactating ewes that are six or more weeks apart in their lactation curves, which thoroughly complicates your feeding management because some ewes will be just reaching peak lactation while other ewes will be at the weaning stage. Even your lamb marketing is affected, because your lamb weights will not be uniform. What a mess.

So how do you efficiently manage such a flock? Long hours in the barn and lots of guesswork about nutrition — with ewes that are too fat or too thin, and lamb weights that vary all over the map. Wouldn't it be simpler to watch your rams carefully at breeding time for aberrant behavior, or to use marking harnesses or brisket paint, or even to use a couple of extra rams? On the other hand, you can just go to that airport ticket counter and buy a ticket to Tahiti . . .

First Published: July 1994

Author's Note: Every year I hear the about this issue or some variation of it. Except from commercial producers who use a lot of extra ram power.

Protein and the Pregnant Ewe

The standard recommendation for feeding twin-bearing ewes in late pregnancy — i.e., feed 0.5–1.0 lb of grain during the last four weeks of gestation — is designed to prevent ketosis. High-producing ewes need this extra energy. Although grains may be low in protein, we don't worry about it as long as we also feed a reasonable quality hay. But we may be wrong. The times, as they say, they are a-changin'.

All livestock fetuses share the same growth pattern: they grow slowly at first as the mother builds placenta and other uterine tissues, but during the last trimester of pregnancy, the fetuses get serious about growth and put on nearly 70% of their final weight. The mother must supply nutrients for that growth, so she increases feed consumption and if necessary, mobilize body fat. In general, a ewe can easily provide enough nutrients to her single fetus, but twins can put a severe nutritional strain on the ewe. And triplets are even worse.

The problem is that converting body fat into usable energy incurs a risk of ketosis in sheep. Thin ewes tend to convert some products of body fat into ketones, which can build up in the blood and cause metabolic problems. Once they become ketotic, ewes also go off feed, which reduces their energy intake and makes the situation worse. Unless the lambs somehow hurry their birth date, a ketotic ewe and her fetuses will die.

To prevent ketosis, the NRC (National Research Council) recommends that ewes carrying multiple lambs should receive some extra energy during late pregnancy. They recommend a diet with 66% TDN and 11% crude protein — higher than their early/mid-pregnancy diet of 55% TDN and 9% protein. For example, if a farmer feeds mediocre grass-clover hay (58% TDN, 12% crude protein) and adds 1.5 pounds of corn (88% TDN, 10% protein), and his 175-pound ewe eats dry matter at 2.6% of her body weight, the final diet of 1.5

pounds of corn plus 3.6 pounds of hay would contain 67% TDN and 11.4% protein, which is pretty close to NRC requirements.

But is this enough?

Alan Bell and his colleagues at Cornell University recently published a paper in the Journal of Animal Science (March 1997) that says no, this is not enough. They fed diets of different protein levels to twin-bearing ewes during their last month of pregnancy and then traced the changes in nitrogen accumulation in various organs. Their results were quite surprising.

Their study used 32 Dorset ewes all pregnant with twins. From Day 80 through Day 110 of pregnancy, they fed all the ewes a balanced diet containing 12% crude protein (dry matter basis). The starting point of this study was Day 110. To establish a baseline on Day 110, they slaughtered eight ewes and carefully measured their body compositions. From Day 111 through Day 140, they fed the remaining ewes one of three diets containing either 7.9%, 11.6%, or 15.7% crude protein. On Day 140 they slaughtered the remaining 24 animals and measured everything they could get their hands on — body composition, blood levels of metabolites, fetal and uterine tissues, and so on.

Some results were expected: ewes that were fed the low protein diet (7.9%) did not do as well as the other animals. Compared to ewes on the higher protein diets, ewes on the low protein diet gained 50% less, they lost more body condition, their lambs (combined weight of twins) weighed 13% less, their uterine tissues weighed 14% less, their livers weighed 18% less, and the mammary glands weighed 44% less. These results seem like a good demonstration that feeding less protein than the NRC recommendation during that last month of pregnancy is not a great idea.

But the researchers also observed some *unexpected* results, particularly on how the high protein diet affected nitrogen levels in the muscle. As the dietary protein level increased from low to medium to high, nitrogen accumulation in the muscles did not level off as expected at the medium protein level (which is the recommended level). Muscle nitrogen continued to accumulate as ewes consumed protein above their recommended amounts.

The data showed that the *low* level of dietary protein forced the ewes to shift nitrogen away from their muscles to support uterine and fetal growth. The *medium* level of dietary protein reduced this shift but did not completely eliminate it. The *high* level of dietary protein, however, actually allowed ewes to *increase* the amount of nitrogen in their muscles.

Translation: muscle tissue was acting as a reserve bank account for nitrogen. When the pregnant ewes did not receive enough nitrogen from their diets, they made up the difference to their fetuses by mobilizing some nitrogen from their own muscles, and to a lesser extent from other organs. This information is new. Although it's been long known that ewes can shift their *energy* reserves — mobilizing body fat is the fundamental cause of ketosis in under-conditioned ewes — it's not been known that they can also do this with *nitrogen.*

Metabolically, energy and nitrogen are two different things. Extra energy is routinely stored as fat, but in mammals, extra nitrogen has *no* dedicated storage molecules. Nearly all the nitrogen in livestock is locked up in proteins, which are large molecules with important functional or structural roles. In livestock tissues, therefore, nitrogen equates to protein. Alan Bell's study showed that twin-bearing ewes that were fed recommended levels of dietary protein still mobilized muscle protein to support their fetuses, but when those ewes were fed extra dietary protein, they changed gears and put that protein back into the muscles.

Whoa! That's a lot of biochemistry. How does this affect the shepherd who is trying to balance rations and feed pregnant ewes?

Well, feeding the recommended level of protein during late gestation resulted in good udder development, fetal weight, and body condition. However, feeding a *higher* level of protein allowed twin-bearing ewes to build up muscle reserves just before going into lactation. Since this study showed that muscle protein was *labile* (scientificese for movable) during pregnancy, it follows that protein may also be labile during lactation, which means that it could be used to produce milk. Which could be fairly important for ewes trying to rear twins or triplets.

What does this all mean? Bottom line: during the last month of gestation, I wouldn't just supplement 10% corn to ewes carrying twins or triplets. I'd think about using a 16% supplement, or in our area of the Pacific Northwest, cull peas — which routinely contain more than 20% protein. Because a little extra protein can go a long way.

First Published: November 1997

Author's Note: This research highlights one of the industry's quiet secrets: that people may know and respect the official reference tables of nutrient requirements, but that doesn't mean that they actually follow them under field conditions. Alan Bell's research is the type of study that sometimes translates to changes in attitude and professional recommendations which cannot be traced to the formal numbers published in reference books.

Nail Polish Remover?

Let's say that you have a flock of 300 mature ewes two weeks from lambing. Most are in reasonable body condition, probably carrying 180% lamb crop. You're feeding them good quality alfalfa hay and free-choice minerals. But one morning as you walk through the flock, you notice three ewes standing off to the side, kind of listless, not eating much, and one is particularly wobbly on her legs. Okay, what's wrong? Automatically, you think "ketosis" and reach for the drench gun containing propylene glycol. That's a reasonable choice, so let's discuss it.

Ketosis — it's *the* classic nutritional problem in late gestation that all shepherds learn about in lambing clinics. It can occur in ewes carrying twins and triplets, usually in underconditioned animals but sometimes in obese animals. Ketosis is essentially caused by the lack of glucose during the last few weeks prior to lambing, when the fetuses are growing rapidly and need lots of glucose, but fetal volume may reduce the amount of feed a ewe can consume. This syndrome is universal, so it has many names, including *pregnancy toxemia*, *twinning disease*, and in New Zealand, it's called *sleepy sickness*.

Ewes suffering from ketosis definitely look sick. They appear listless, ears droopy, possibly trembling. They hang back from the rest of the flock, reduce their feed intake, and sometimes stand in a corner pressing their head against a wall. Eventually they go down and may refuse to rise again. Their breath has the distinct sharp odor of acetone.

Acetone? Well, acetone is the result of the ewe's biochemistry, and this story, although a little convoluted, is *very* interesting. As the ewe desperately tries to support her fetuses in late gestation, she mobilizes fat off her back, which means that she sends hormonal signals to her fat reserves which then export fat molecules into the blood. Those molecules are transported to the liver where they are enzymatically broken down into smaller molecules that are metabolized for energy, especially for the production of glucose. So far, so good. But this process also produces a 4-carbon byproduct compound called *acetoacetate*, some of which is then transformed into a related 4-carbon compound called *beta-hydroxybuterate* (also occasionally printed as *β-hydroxybuterate)*. Both compounds enter the blood and are carried to tissues around the body, where they can be used for energy. Meanwhile, a small portion of the acetoacetate slowly degrades into *acetone*.

However — and there is *always* a however — *in some ewes* the production of these byproducts outstrips the capacity of peripheral tissues to use them, and their blood levels rise. These three compounds — acetoacetate, β-hydroxybuterate, and acetone — are called *ketones* (or *ketone bodies*, although this label has nothing to do with dead chemical bodies), and if ketone levels in the blood become too high, the ewe suffers from ketosis.

During ketosis, some of this excess acetone ends up in the urine and in the rumen, and then in the gas released from the rumen. You know that odor of acetone. It's the same chemical that you can buy in a hardware store as an organic solvent. It's also the active ingredient in some nail polish removers.

Classically, ewes in poor body condition are most susceptible to ketosis (*Starvation Ketosis*). This makes biochemical sense, really, because these animals are caught in a metabolic bind — trying to support multiple fetuses and also regain body condition at the same time. But paradoxically, ewes in excessive body condition are also susceptible to ketosis (*Estate Ketosis*). Although these ewes have lots of body fat and are obviously eating well, their livers still can produce high levels of ketones, although the precise mechanism is unclear.

So what can we do? On the ambulatory level, we can drench a ketotic ewe with something that she can metabolize quickly into glucose — either propylene glycol or glycerol (4–8 ounces per day). Both compounds pass through the rumen unaffected by the microbes and are absorbed quickly into the blood and then transported to the liver. Liver cells then transform these compounds into the glucose which the ewe needs to prevent the buildup of more ketones. And we continue this treatment until the ewe goes back on feed or lambs or dies. Advanced cases may respond to intravenous glucose. Ironically, feeding more corn or molasses or even sugar water is not as effective quickly, because the rumen microbes convert some of that energy into compounds that the ewe cannot use to make more glucose.

On a prevention level, we need to monitor body condition during pregnancy. One nutritional guideline for the last six weeks of gestation, for ewes carrying twins or triplets, and especially for ewes with condition scores lower than 3.0, is to feed 0.25–0.75 pounds of grain (corn, barley, etc.) in addition to their forage. We are not trying to fatten animals, just trying to provide them with enough extra energy to prevent fat mobilization.

Obese ewes, however, are a problem of a different color. These are animals with condition scores of 4.5 or 5.0 on a 5-point scale. We all know some

— they like to eat. But here's the dilemma: on one hand, feeding them extra grain will get them fatter (if that were possible) which makes them *even more* susceptible to estate ketosis. On the other hand, reducing their energy intake during late gestation creates a declining plane of nutrition which will *also increase the risk* of ketosis. No good answer here — just one principle: don't have fat ewes prior to lambing. Avoid this problem by controlling body condition in the 30 weeks *before* late pregnancy.

What about other species of livestock? Ketosis is well-known in the dairy industry — it can easily affect 10% or more of the highest producing cows. The symptoms in cattle are analogous to sheep, with a similar increase in blood ketone levels and the trademark smell of acetone on the breath, but the consequences in cattle are vastly different.

A dairy cow only suffers ketosis *after* she calves, when her milk production is beginning to rise quickly. Nutritionally, she is in negative energy balance during this period, and she must mobilize some fat. But ketosis is not deadly in dairy cattle. It's actually self-correcting. When a ketotic cow backs off feed, her milk production declines, and therefore her demand for energy declines, along with the metabolic stress that causes the problem. Her ketosis subsides and the cow eventually goes back onto feed again. The dairy farmer experiences some financial loss, of course, but the cow remains alive to produce more milk.

In contrast, ewes have no such luxury — ketosis in ewes usually occurs *before* lambing. Since the cause is related to massive fetal growth during late gestation, there is no way of reducing the stress except by lambing. Or by aborting the lambs. Or by death of the ewe. And when a ketotic ewe backs off feed, she only makes the stress *worse*. Her metabolism quickly spirals down towards a fatal conclusion.

So we try hard to avoid ketosis in our ewes. Not always a simple task. And ketosis may not always be simple either. Another syndrome — low blood calcium — can also affect ewes in late gestation, and its symptoms are similar to ketosis. Hmmm. Let's talk about it next month.

First Published: March 2002

Author's Note: Metabolic problems are not fun, either for the rancher or for the animal. And sometimes their treatments are not intuitive. For ketosis, which is a syndrome caused by a lack of nutritional energy, we don't feed an obvious source of energy (corn or barley), but rather we drench the animal with a compound used to manufacture the explosive nitroglycerin (glycerol) or a compound commonly found in toothpaste (propylene glycol). And drenching with nail polish remover is never recommended. Life has its interesting turns.

By the way, for the "next month" article listed in the last sentence, look at the chapter called "For More Than Strong Bones."

Percent of What?

A few shepherds and I were recently discussing the lambing rates of some local ranches, and we ran into a problem. We discovered that, although we were using the same words about lambing rate, each of us meant different things.

Lambing Rate — you'd think that such a common sheep industry term would be universally understood, right? Wrong. During our conversation, it became painfully clear that we all held different definitions of lambing rate. Our discussion quickly degenerated into a confusing babel of conflicting phrases, distressingly reminiscent of the famous Abbott & Costello skit "Who's On First?"

Why? Because the numerical percentage that we report as "Lambing Rate" is really a fraction — a ratio between two numbers. The top number of this fraction (the *numerator*) is the number of lambs, and the bottom number of this fraction (the *denominator*) is the number of ewes, but *which* lambs and *which* ewes? The devil is in the details.

Let's first look at the numerator. Many producers define this number to include *all* lambs — i.e., the sum of *every* lamb born alive, dead, and questionably in-between (for a while). But not everyone agrees on this. The numerator can also mean *only* live lambs, or *only* lambs that were tailed, or *only* lambs that were marketed. For the purposes of this article, let's identify these numerator options as N1, N2, N3, and N4, respectively.

The denominator can also come in a couple of flavors. Generally, producers define the denominator to include *all* ewes exposed to rams. Alternative definitions, however, may include *only* those ewes who actually lambed, or *only* those ewes who raised lambs to weaning. There may be other definitions, but these are the most common. Let's identify these three denominator options respectively as D1, D2, and D3.

Even this quick list gives us four possible numerators and three possible denominators — a total of *twelve* possible combinations. And for each one, someone, somewhere, has defined it as their lambing rate (LR). So, in the interest of clarity, let's examine some of the most widely used definitions.

Probably the most common LR definition is *total number of lambs born* — alive and dead — *for all ewes exposed to the rams.* Using our shorthand, this would be an N1-D1 combination. "All lambs" includes those lambs born dead. We must be careful here. Should we also include a late-term abortion if we discover the expelled fetus? Yes, I would, because this LR value tries to identify the number of successful fertilizations for the flock, and an aborted fetus still represents a successful fertilization. Including all ewes in the denominator emphasizes the financial aspect of lambing — since we must feed barren ewes as well as fertile ewes. This N1-D1 definition has one potential weakness, however — it doesn't account for any barren animals sold before lambing time.

The next most common LR definition is probably the N1-D2 combination — all lambs born *per ewe lambing.* Here, we are concentrating *only on ewes that produce lambs.* This LR measures *prolificacy,* since it reflects the number of eggs produced by the ewe *plus* her ability to carry those embryos to term, and it ignores open, unproductive animals. The resulting LR, sometimes also called the *Drop Rate,* gives the highest value of our array of LR possibilities because it maximizes the numerator compared to the denominator. Many producers like this definition because it is particularly good for bragging rights.

Some shepherds prefer a different LR value — the N2-D2 combination — number of *live lambs* per ewe lambing. This LR combines the fecundity perspective with the ability of the ewe to produce a saleable product (assuming that there is not much market for dead and aborted lambs). Intensive sheep operations may use this LR, because it presumes that someone is present at lambing to observe weak lambs before they die. Since a high percentage of lamb mortality occurs within the first twelve hours after birth, shepherds have developed many heroic techniques for saving these weak animals. This LR definition is actually kind of a hybrid value because it reflects the interaction of the ewe's ability to produce lambs and the shepherd's ability to keep those lambs alive, at least through the initial few hours. This LR, however, does not measure mothering ability, which is a trait that fully develops after lambing is completed.

Large range operations may use a fourth LR definition — the N3-D1 combination — *the number of lambs processed (banded, docked, castrated)* per all ewes in the flock. This LR value represents a kind of global measurement, a long-view approach to lambing, where large numbers of ewes lamb without minute-to-minute observation or assistance. These ranches let their ewes lamb more or less on their own, and then periodically gather the animals for processing. This LR value will be lower than an N1-D2 combination, because the numerator is always reduced by lamb deaths at birth, while the denominator includes open ewes. But this LR also puts strong emphasis on selecting the practical traits of easy-lambing, lamb viability, good mothering ability, and good predator control, as well as the economic necessity of reducing labor. Although traditional in some regions, calling this value a "Lambing Rate" is a bit misleading. Perhaps we should call it a *Marking Rate* or *Banding Rate*, since the actual lambing events may have occurred weeks earlier.

And if these definitions aren't confusing enough, I need to add two more ingredients into this stew: (1) ewe lambs and (2) official reports.

The first one is ewe lambs. Ewe lambs inevitably show lower fecundity and more open animals than adult ewes. Including ewe lambs in a flock's LR masks the true productivity of the adult ewes. Therefore, should an LR include ewe lambs? If yes, then how can we compare LR values from flocks that breed ewe lambs with flocks that don't? What about flocks in expansion mode that retain high numbers of ewe lambs? How can those flocks compare their LR values to flocks containing smaller proportions of ewe lambs or even to themselves in later years when they are not in expansion mode? The best statistical answer here is to report *two* LR values (for whichever definition) — one for the adult flock and one for the ewe lambs.

And then there are those pesky official reports — statistical documents published by various government agencies and private organizations. Those reports usually list a national (or state or provincial) LR. But how do these organizations gather their data? By sending questionnaires to producers, who as we've seen, are not exactly unanimous or consistent in their definition of LR. And when the field data comes back to these organizations, their well-meaning office folks make arbitrary decisions about which numbers to use. I suspect that pure science may not be at the top of their priorities.

So the next time you read those reports or hear someone mention lambing rate, you might ask, "Percent of what?" And be prepared for a surprising answer, because sometimes, we don't know who is on first.

First Published: January 2004

Author's Note: Are you thinking — yes, there may be confusion in the field, but so what? Isn't this confusion only a problem of semantics? No, it's a lot more than semantics. Each of these ratios tells us different information about an operation. If a shepherd does not recognize the importance of all these different definitions of lambing rate, then she won't use them effectively for analysis and decision-making. I've seen this happen, with sad results.

Pellets in the Jug

Feeding ewes in a typical lambing barn: some hay; a little bit of grain — next jug — some hay; a little bit of grain — next jug — some hay; a little bit of grain, etc., etc. Oh yes, now the water: fill this bucket; lift it over the panels — fill the next bucket, lift it over the panels, etc., etc. And now the bedding — first we have to clean out the jug before the next ewe and haul that bedding away, as well as remove all the loose hay that the previous occupant inconsiderately dumped into the straw, then we move to the next jug and . . .

Those lambing jugs . . . so little time, so much to do. How can we use our time better during lambing? Let's discuss feeding ewes in jugs. I'll be more precise: what does a newly-lambed ewe actually *need* while she's in the lambing jug? I'm not asking what everyone *feeds* — we could take a poll, and I'm sure the answers would be fascinating — a veritable raft of homegrown remedies. For the moment though, here's the situation from a nutritionist's point of view . . .

Let's assume that in our lambing jugs we have 154-pound ewes (= 70 kilograms. Nutrient requirement tables are initially published in metric units. Apparently, all scientific ewes are metric). Our ewes are in fair body condition and are not suffering from ketosis or other obvious problems. Some have just lambed twins, some singles. What should we feed them?

So I go to my trusty *SID Sheep Production Handbook*, the nutrition chapter, and look up the nutrient requirements for lactating ewes.

For ewes in early lactation, the Nutrient Requirements table is divided into two categories: (1) ewes with twins and (2) ewes with singles. Early lactation is defined as the first 6–8 weeks after lambing. Uh, oh — I immediately see a problem. Consider what happens to a ewe during lactation: milk production begins at a certain level, rises to a peak at 3–4 weeks, and then begins to drop off. Ewes milking twins produce only 40% more milk than ewes milking singles, more or less. The nutritional requirements listed in this table actually represent a comfortable middle ground. These requirements were designed to provide enough nutrients to the ewe to prevent extreme body condition loss while she produced milk for one or two voraciously hungry lambs. These nutrient requirements were designed for the long haul — to feed ewes across their entire early lactation period of 6–8 weeks.

But the ewes in our jugs are not ewes in the middle of their early-lactation period. Our ewes are simply newly-lambed ewes in jugs *during the first 1–3 days of lactation*, when their milk is just beginning to flow. Of course each ewe has one or two lambs, but if you think about it, how much milk do those feeble lambs actually consume? Remember, when those ewes were brought into the jugs, their udders were full of milk and their rumens were full of feed (hopefully). That feed is still merrily fermenting and digesting — you might consider this system as a nutritional pipeline — and it will take *many days* to use up all the protein and energy already in that pipeline.

So why would we try to feed these newly-lambed ewes using the same recommendations designed for ewes milking heavily during the middle of their early lactation period? Those reference tables call for high nutrition levels to support lots of milk over many weeks. If we consider this carefully, it's kind of like digging a post hole with a front-end loader. In the end, there will definitely be someplace to put the post, but a more precise or convenient tool would do a better job. The bottom line for those of us in the lambing barn is that, during lambing, these nutrient recommendations should be tempered with good judgment.

Basically, it makes sense to use a ration that is easy to feed, easy to obtain, and nutritionally well-balanced. Price is not too important, within reason. After all, you won't use this ration for very long, and other advantages like convenience and time efficiency could outweigh its additional costs (unless you figure that your labor isn't worth anything). Therefore, why should we feed the traditional ration of loose hay and grain? Loose hay is certainly not convenient, unless you LIKE dust and wastage and hauling. And grain is usually not necessary for these ewes, nutritionally.

While some people like to feed grain to keep their ewes calm, that's a matter of judgment. My ewes are anything but calm when they sense the approach of a grain bucket. And the resulting large udders may be more susceptible to problems of abrasions and mastitis, particularly in many barnyard environments. Of course, sheep like grain — so what else is new? Sometimes I think that grain is fed to keep the producers happy, as well as the sheep.

Back to our feeding situation: why not use a pelleted legume forage as the sole feedstuff in the jug — like alfalfa pellets or clover pellets? These pellets are certainly convenient, easy to obtain, high in calcium and protein, with no waste. Each day, you just dump a couple of scoops into a tray that conveniently hooks onto the jug panel. A person could walk down a row of jugs and feed them all in 30 seconds. Not bad.

Pellet size is not critical, but the pellets should hold together well, without fines or dust, and they should contain at least 16% protein. Cost? Compare the cost of no-waste pellets to the *real* costs of wastage and haulage of loose hay, not to mention your time. This cost comparison is for a short period, of course. Once the ewes leave the jugs for the mixing pen, they go onto your normal feed.

If you combine this feeding strategy with slatted floors, without bedding, you would have nothing to haul away. Careful, this could create a problem. You could end up having too much time on your hands during lambing. With all that extra time, you might be tempted to do something really outlandish, like get more ewes.

First Published: January 1994

Author's Note: This was written long before the publication of the 2007 edition of the NRC Sheep Nutrient Requirements. Although some of the nutrient requirement values may have changed, the number of hours in a day has not. The principles are the same. We generally spend way too much time at lambing doing things we can do far more efficiently. Feeding ewes in jugs is something that we can easily change.

Where Has All the Fiber Gone?

For years I've recommended a very simple diet for ewes in their lambing jugs: a few pounds of alfalfa pellets in a tray hanging on the side of the pen. Nutritious, convenient, and waste-free. But at various meetings over the years, shepherds have raised two reasonable concerns about this technique, and I'd like to address them here because they both involve basic concepts about forages and fiber.

Concern #1: A sudden switch to pellets will reduce digestion efficiency and cause problems. Reducing "digestion efficiency" means reducing feed digestibility enough to cut milk production drastically or cause health problems with the ewes and lambs. This is a complex problem, so let's examine this situation thoroughly.

Modern nutrition laboratories analyze the dry matter of a feedstuff into two main fractions: cell contents and cell walls. The cell contents include the non-fibrous, soluble, and liquid stuff inside cells. These cell contents are essentially 100% digestible. Pelleting doesn't change this digestibility. The dry matter of medium-quality alfalfa contains approximately 54% cell contents, and none of it is affected by pelleting.

The cell wall fraction of a feedstuff is called *NDF* (*Neutral Detergent Fiber*). NDF is not a simple substance — it's actually a complex assortment of different fiber compounds that combine to form the rigid structure of plant cell walls. NDF contains three basic types of fibers: cellulose, hemicellulose, and lignin. The common term *ADF* (*Acid Detergent Fiber*) is a subset of NDF and is composed of only two of these fibers: cellulose and lignin.

Medium-quality alfalfa hay typically contains 26% cellulose, 10% hemicellulose, and 9% lignin. These three fibers equal the 46% NDF fraction of the plant (the 1% difference from 46% is due to other very minor components). We'll take a look at each type of fiber separately. Lignin is, for all practical purposes, totally *indigestible*, which means that *all of it* passes unchanged completely through the gastrointestinal tract into the manure. Numerically, this means that its digestibility is zero. Pelleting doesn't alter this digestibility, since it's rather difficult to reduce a zero any further. Therefore, any changes in "digestion efficiency" cannot occur in this lignin fraction.

Cellulose and hemicellulose, on the other hand, are fibers that are *potentially* digestible. Their digestibilities depend mostly on rumen fermentation, as in how much can the rumen microbes ferment these fibers and how long do these fibers remain in the rumen to be fermented? Any changes in "digestion efficiency" would occur in these types of fibers.

The nutritional principle here is fairly straightforward: fiber digestibility is defined by its rumen fermentation, and the longer a fiber remains in the rumen, the more it is fermented.

This is where pelleting may have a real effect. Feeding pellets usually increases feed intake. Ewes generally eat more pelleted alfalfa than long hay alfalfa. Increased intake reduces the retention time in the rumen and therefore reduces the digestibilities of the cellulose and hemicellulose. But how much? Ah, that is the question.

Let's see if these changes have real biological meaning for these ewes. The total amount of *potentially digestible fiber* is the sum of the cellulose and hemicellulose fractions of the feed, which for this alfalfa would be 36% of the total dry matter (= 26% + 10%). Any changes in "digestion efficiency" would have to occur in that 36% fraction.

If the normal digestibility of this fiber (when fed as long hay alfalfa) were 60%, then this fiber would provide 21.6% digestible nutrients (= 60% of 36%). If pelleting reduces rumen retention time enough to shorten fiber fermentation by, say 25%, that would mean that instead of a 60% digestibility, the fiber in alfalfa pellets would only have a digestibility of 45% (= 60% less the reduction of 25%). In pelleted alfalfa, the 36% cellulose/hemicellulose fraction would yield only 16.2% digestible nutrients (= 45% of 36%). Still with me?

This means that pelleting alfalfa could reduce the amount of absorbed nutrients (from the cellulose/hemicellulose fraction) from 21.6% to 16.2%. In other words, the total digestibility of the entire pellet would be lowered by 5.4% (= 21.6 – 16.2).

That's not very much when you consider that this would occur *only* during the 1–3 days when ewes are in the jugs. After that brief period, the ewes go into mixing pens where they are fed their regular early lactation ration of grain and forage. So . . . a potential nutritional loss of only 5.4% for less than three days is hardly measurable and is biologically not significant.

Concern #2: Ewes can't adjust to the alfalfa pellets quickly and therefore will suffer from digestive problems. Well, the obvious question is, adjust from what? Let's examine how ewes are fed before they go into a jug.

Ewes in late pregnancy are usually kept in a drop flock or gathered near a barn. Their diet typically consists of hay or silage, plus some grain. Most current recommendations suggest that ewes with multiple fetuses should receive a daily supplement of 0.5–1.0 pound of grain (corn, barley, etc.) during the last four weeks of pregnancy. The grain provides enough extra energy to prevent the metabolic problem of *ketosis*. Ewes on pasture — particularly those lambing late in the spring — may or may not receive any grain during late pregnancy.

All these late-gestation diets, however, are based on forage, which is just another way of saying that they are all relatively high in fiber. For example, medium-quality alfalfa hay contains 46% NDF and 35% ADF, early-bloom grass hay contains 61% NDF and 34% ADF, and young spring grass contains 55% NDF and 31% ADF (all on a DM basis). Silages would have similar fiber levels. In contrast, shelled corn contains only 9% NDF and 3% ADF.

Some people think that alfalfa pellets are somehow very different from the alfalfa hay from which they are made. Not really. Typical medium-quality alfalfa pellets (17% protein) contain 46% NDF and 35% ADF — the same as the hay. The only major difference between pellet and hay is the physical *form* — the pelleting process smushes fiber into smaller pieces ("smush" incidentally, is a technical term). Sure, pellets may differ from long hay in consumption level, fermentation time, salivation amount, heat damage, etc. — but these factors don't alter the overriding principle that alfalfa pellets contain as much fiber as other forages. Nutritional problems during diet transitions are usually caused by *major* differences in fiber content between the two feeds — especially going from a high-fiber diet to a low-fiber diet. Feeding alfalfa pellets to ewes in the jug is not a major switch from a late gestation diet — it's really just changing the form of the fiber.

So, where has all the fiber gone? Into the ewes, of course.

First Published: January 1999

Author's Note: At the end of a workshop presentation, I am often faced with a barrage of questions. There are always some questions that imply that whatever I said won't work because, uh, it just won't. But there are also some questions that are good and reasonable, that require thoughtful responses. This article is an extended complement to the shepherds who expressed these specific concerns. In workshops, my standup, on-the-spot answers tend to be short, but here in the quiet of these written paragraphs, I can take some time to expand on my responses and ruminate about the details.

Choices, Choices – Not Enough Milk

I once heard a story about the fellow who drove from Chicago to Iowa. When he first started his car in Chicago, he noticed that the dashboard engine warning light was on. But he didn't have time to fool with any repairs, so he did the next best thing — he taped a piece of duct tape over the light and drove west. He arrived in Iowa all right, but once he turned the engine off, the car never started again.

Feeding ewes in early lactation can be just like that.

We know that ewes in early lactation need good nutrition, but how would we know if anything is "broken?" Here's an example: let's say that we're nearly finished lambing, and lots of our ewes are raising twins. We look suspiciously at that barn full of hay: how good *is* that hay, really? And if the hay lacks something — what can we do about it? Maybe we should just feed it out and see what happens. Duct tape, anyone?

Let's approach this problem in three stages. The first step is to gauge what the ewes need. The second step is to determine if our hay meets those needs. And the third step, when necessary, is to consider some alternatives.

Step #1: We'll use our trusty *SID Sheep Production Handbook* as our reference, particularly the reference tables in its nutrition chapter. Let's look up the daily nutrient requirements for ewes suckling twins during the first 6–8 weeks of lactation. All our ewes weigh exactly 176 pounds, of course, because the reference table contains a convenient row for that weight. We see that our lactating ewe requires 4.3 pounds of TDN (energy) and 0.96 pounds of crude protein. We'll ignore her mineral requirements for now, because these are easily fulfilled by offering an appropriate mix of a trace mineral salt, limestone, and/or dical (dicalcium phosphate).

Table #4 also lists dry matter intake (DMI) at 6.6 pounds. Remember, however, that DMI is *not* a requirement. *DMI is only an estimate of intake based on the assumption of a high-quality ration.* If the ration is of lower quality, then intake will generally be lower.

In the field, a good practical recommendation for feeding a twin-rearing ewe is to feed two pounds of 16% supplement and all the good alfalfa hay that she can eat. The typical 16% grain supplement contains 90% dry matter. Therefore, when we convert these values to a dry matter basis, *a 16% supplement really contains 90% TDN and 17.8% crude protein* (= 16 ÷ 0.90). Thus, two pounds of supplement equals 1.8 pounds of dry matter, which supplies 1.6 pounds of TDN and 0.32 pounds of protein.

After we feed the 2 pounds of supplement to the ewe, our hay must still supply the remaining nutrients — i.e., 2.7 pounds of TDN and 0.64 pounds of protein. If our girl can eat 4.8 pounds of this hay, its quality would only need to be 56% TDN and 13% protein. Still with me?

Step #2: Is our hay good enough? We can't just guess; we must take a forage test to obtain real numbers. But a forage test is an especially easy procedure in this article. In the third sentence, we can send our sample to the lab, and by the fifth sentence, we receive the report. The test report lists our hay at 58% TDN and 12% crude protein. Seems fairly close.

But . . . and there's always a "but" . . . will our ewe actually *consume* 4.8 pounds of this hay? Probably not, unless she is part-Hereford. Our hay isn't exactly leafy green alfalfa — which is why we were concerned in the first place. There are quite a few stems in those bales. The forage test lists an NDF (Neutral Detergent Fiber) value of 65%, which confirms our suspicions. High levels of fiber reduce hay intake because of rumen fill and a slower rate of passage.

Realistically, we may expect our ewe to consume only 3.5 pounds of that hay dry matter, not 4.8 pounds. This lower amount would only provide 2.0 pounds of TDN and 0.42 pounds of protein.

Now we can see the real problem: a shortfall of 0.7 pounds of TDN and also 0.22 pounds of protein. This translates to lower milk production and slower lamb growth.

Step #3: Now what? Should we run out and buy 40 tons of leafy alfalfa hay? Hmmm. In some parts of the country, good alfalfa costs more than $100 per ton. At that price, even Microsoft would have to get a bank loan first.

Couldn't we simply feed extra grain to the ewes? I wouldn't. The ewes, of course, would gladly gobble up any grain I fed them. But this extra grain would *replace* some of the hay rather than completely add to it. This may not be good economics.

One possible strategy would be to feed extra hay and deliberately accept a higher level of wastage. This will allow the ewes to pick through the forage and choose the higher-quality leaves — in effect *increasing* the hay's nutritional value. The economics of this choice depends on our hay inventory, our labor, and the relative prices of replacement forages.

What about pelleting the hay? Pelleting, unfortunately, would actually *decrease* the hay's nutritional value because pellets force ewes to consume *all* parts of the hay, including the fibrous stems. Pelleting mediocre hay may be a good strategy for controlling wastage, but it's a bad strategy for providing extra nutrition to lactating ewes.

Alternatively, we can change our priorities and focus on the lambs rather than the ewes — we can install a creep feeder to give the lambs extra energy and protein. The downside is that, in some housing situations, creep feeders can also increase the risk of infectious diseases and foot problems.

So here's my favorite option: we do nothing differently, at least at first. We accept a slower lamb growth during lactation, but then we simply end lactation early. In other words, we direct our nutrition to where it is needed most, into the older lambs which represent our real marketable crop. We wean the lambs early at 60 days onto high-quality feed — grain or spring pasture, whichever makes economic sense. Remember that young spring pasture contains as much digestible energy as oats and considerably more protein. Pastured lambs may not gain as fast as lambs fed corn and peas, or look as fat, but the cost of their gain will be considerably cheaper.

By the way, I'm *not* going to throw out my duct tape quite yet — I may need it to tape down the baling twine which holds my fences together.

First Published: February 1996

Author's Note: The specific nutritional requirements mentioned in this article are based on the 1985 edition of the NRC *Nutrient Requirements of Sheep*. The NRC recently published an updated edition of these tables, but that doesn't alter this problem or the options I've presented. The principles of matching nutrient requirements with nutritional quality and dry matter intake do not change. Our options do not change. And our desire to use duct tape also does not change.

Grain on Grass – Let's Do the Numbers

Okay, raise your hand — how many of you have supplemented grain to animals while they were grazing on pasture and were disappointed with the response? Don't be shy; keep your hands up. Well, you're not alone. Twice during the past year I've read scientific papers that reported the same thing, and those researchers were not only disappointed with the results but were also puzzled. After all, why wouldn't extra grain provide enough surplus energy to overcome intake problems and increase daily gain or milk production?

Because you wouldn't expect it.

First, the standard answer. In every university course called *Livestock Nutrition 101*, there is a session when the instructor lectures about the practical issues of supplementing grain to grazing animals. Essentially, his message boils down to the precept that grain will "replace" some of the forage and therefore will not provide as much extra nutrition as you'd expect. The instructor continues with an example: if a ewe was consuming 6 lb of forage, adding 2 pounds of corn will not simply boost her intake to 8 lb. Most of the corn will *replace* some of the forage, and total feed intake will rise only 0.5 lb or so. Since the TDN value of corn is 88% (all nutritional values in this article are on a dry matter basis), and the TDN value of the forage is, say, 65%, the net effect of all this supplement is only a modest increase in nutrient intake — and certainly not as much as the 1.6 lb of TDN that you'd expect from 2 lb of corn (90% DM at 88% TDN). The students dutifully write this down and perhaps ask a question or two, and then the instructor moves on to the next topic, maybe something about the effects of chewing gum on hippopotamus growth or whatever.

Elementary, my dear Watson, elementary.

Now, let's move beyond this simplistic explanation and look at grain supplementation in more depth. Grain doesn't just "replace" forage. Grain also profoundly changes the rumen environment, and these changes can sometimes offset much of the extra energy supplied by the grain. Nutrition textbooks typically list this phenomenon as the *Associative Effects*, but let's see what these effects really mean. Oh yes, you can put your hands down now.

We need to start with four assumptions: (1) the supplement consists of corn or barley or a multi-grain mixture and does not contain any added buffer such as sodium bicarbonate, (2) a significant amount of grain is offered, (3) the pasture is reasonable quality with a TDN value of 65%, and (4) the grain is offered only once each day, which is the typical procedure on most farms. These assumptions, of course, imply the following: that the grain supplement is primarily starch, that the supplement does not contain lots of salt to limit intake, and that the supplement is consumed rapidly. The last assumption is fairly obvious to anyone who has ever fed corn on pasture. Aside from protecting yourself against being trampled, you'll observe that the animals will nearly always *inhale* the supplement — they gobble it up as fast as their mouths can move. No dainty manners here. In all the years of feeding supplements, I've *never* seen animals step back to save some grain for a future late-night snack.

So here's what happens when this grain is supplemented to grazing ruminants: The starch in the grain enters the rumen and ferments at a very fast rate, much faster than fiber. The rumen bacteria that ferment this starch produce end-product acids (*VFAs — volatile fatty acids*) so quickly that these acids overcome some of the buffering capacity of the rumen, driving down the rumen pH from its normal level of 6.2–6.5 to less than 5.8, at least for a few hours each day.

The lower rumen pH causes problems for the species of bacteria that ferment fiber. The lower rumen pH reduces their populations and activities, thus slowing down the rate of fiber digestion. Because the undigested fiber remains longer in the rumen, sensors in the rumen wall alert the animal's neural feedback system that the rumen is still full. Which tells the animal to reduce its feed intake. Since we assume that the animal eats all its supplemental grain, any reduction of feed intake must come from the amount of grazed forage.

Therefore, grain supplementation on pasture results in a lower intake of forage *and also a lower digestibility of that forage*. And for those who are still following me, this effect would be more pronounced with grass than with a legume such as clover or alfalfa. Why? Because grass contains higher levels of potentially-digestible fiber than legumes, and it's the fermentation of this potentially-digestible fiber that is most depressed by the feeding of starch.

Now, let's do the numbers. Our example will be a 154-pound ewe suckling twins in early lactation (using the 1985 NRC Nutrient Requirements for a 70-kilogram ewe). This ewe requires 4.0 lb of TDN to support her milk production and minimize her early-lactation weight loss. If she grazes pasture containing 65% TDN with a daily dry matter intake of 4.0% of her body weight, she would eat 6.16 lb of dry matter (= 4% of 154) containing 4.0 lb of TDN, which nicely meets her requirements.

But . . . let's say that we want to increase milk production or prevent loss of body weight, so we'll offer this ewe a daily supplement of 2 lb of corn (= 1.8 lb of dry matter). Since corn is 88% TDN, this supplement will provide 1.58 lb TDN. And of course, our ewe will gladly eat all the corn quite rapidly.

If we assume that the ewe's dry matter intake will rise slightly — to 6.8 lb — then her forage intake will be 5.0 lb (= 6.8 minus 1.8 of corn). If we *ignore* the associative effects of the starch and assume that the original nutritional value of the forage remains unchanged at 65% TDN, we can calculate that 5.0 lb of forage will provide 3.25 lb TDN (= 65% of 5.0), making a total TDN intake of 4.83 lb — which is a 21% increase of digestible energy intake due to grain supplementation. Hmmm, so far, 21% looks pretty good.

But we can't ignore the associative effects of starch on fiber digestion, can we? Of course not. Therefore, if we accept that associative effects apply to our situation, then we must reduce the TDN value of the forage from 65% to, say, 55%. Now let's *redo* the numbers with this new TDN value.

Our ewe consuming 5.0 lb of this forage will now only receive 2.75 lb of TDN from it (= 55% of 5.0). Adding the 1.58 lb TDN from the corn gives her a total daily intake of 4.33 lb TDN, *which is only 8% above her original energy requirements.* Not exactly something to write home about. In the highly variable world of real-time grazing, a TDN boost of only 8% would be lost in the normal background variation.

Let's put this in perspective: in our example, the supplementary grain provided +0.83 lb TDN when we *didn't* include associative effects in our calculations but only +0.33 lb TDN when we *did* include them. The difference between these two numbers represents a 60% drop in supplemental TDN from the corn (= 0.5 as a % of 0.83). Which number is correct? Well, how many times have you been disappointed by the performance of grain-supplemented animals on pasture?

There is, of course, an alternative way of looking at this situation: even if the animals didn't perform as well as expected, at least we can be assured that all that corn made them happier.

First Published: September 2004

Author's Note: I've observed many instances where farmers and ranchers were disappointed when they tried to improve production of their grazing animals by supplementing grain. This is a situation where the complexity of ruminant nutrition becomes more than an odd topic discussed in an academic course.

Washy Grass

Winter in Douglas County is upside down. It's early February as I write this, and while the rest of North America is bracing for another arctic storm, Douglas County is basking in a balmy 60° weather, with lots of rain and occasional sunshine. Three inches of bright green grass already cover our pastures. Lambing is nearly finished, and ewes and lambs are moving out into the paddocks. The question is, can the ewes drink enough grass to support lactation?

I'll be more precise: young three-inch grass during our winter (read: early spring in the rest of the country) is highly nutritious, but it also contains lots of moisture. Ewes in early lactation need high levels of energy and protein, especially if they raise twins. Can these young pastures do the job?

Nutritionally, immature grass is almost like a grain. Even though our clovers haven't started to grow yet, these young grasses show TDN values of 70% or more, with crude protein levels over 20%. On a dry matter basis, our forage has the same energy value as oats, with approximately twice as much protein.

On a dry matter basis — what *exactly* do I mean? Well, every feedstuff contains *some* water. Oats, corn, and hay all contain at least 10–12% water; balage contains 50% water; corn silage contains 70% water; and cow's milk contains 87% water. Because water has no nutrients, the only way we can properly compare feeds is to report their nutritional values on a dry matter basis. Converting all values to dry matter (by dividing the original number by the dry matter percentage) levels the playing field. Conversely, feed values that include water are labeled "as fed" basis or "wet" basis. But water can be misleading. For example, on a wet basis, milk contains only 4% protein, which makes it *appear* poorer than ordinary 12% hay.

Western Oregon forages in late spring or on irrigated summer pastures contain 25% dry matter = 75% water. Although this seems like a lot of water, it's typical of living tissue and isn't enough to cause problems.

But things are different for forages growing during the winter and very early spring. These young grass leaves may contain only 17% dry matter = 83% water. Let's see how this may affect our ewes . . .

Consider a typical 154-pound Douglas County ewe raising twins in early lactation. (Why 154 pounds? Because 154 pounds equals 70 kilograms, which is a nice round number listed in metric reference tables.) The National Research Council (NRC) recommends that this ewe needs 4.0 lb of TDN and 0.92 lb crude protein each day. The NRC also suggests that she will eat 6.2 pounds of dry matter, which is 4.0% of her body weight. These requirements translate to a feedstuff containing 65% TDN (= 4.0 ÷ 6.2) and 14.8% crude protein (= 0.92 ÷ 6.2), on a dry matter basis.

Based on our forage values of 70% TDN and 20% crude protein, shouldn't we conclude that these young winter grasses are okay? A ewe consuming 6.2 pounds of this forage dry matter will receive 4.3 lb of TDN and 1.24 pounds of protein . . . comfortably more than she needs. But if this actually happens, then why am I writing this article?

Because the basic assumption is flawed. *The problem is that our ewe won't eat 6.2 pounds of dry matter.* The forage has too much water . . .

Ewes certainly like to eat, but even *they* have a limit. Ewes don't spend more than 12–13 hours each day actively grazing. During each hour of grazing, they take a certain number of bites, and each bite contains only a small amount of grass. To figure a ewe's daily dry matter intake, just multiply the following: (number of hours grazing) *x* (number of bites per hour) *x* (dry matter intake per bite). The problem is this: if the dry matter percentage of the forage goes down, then a ewe must compensate by changing the other two factors. Otherwise, her total nutritional intake also goes down.

Compare a bite of normal grass at 25% dry matter versus a bite of young grass at 17% dry matter. This difference represents 32% less dry matter in a single bite (= 25 – 17, as a percentage of 25. Still with me?). I think that we can confidently say that grazing time will not increase by 32%. Also, I seriously doubt that ewes can increase their grazing speed (bite rate) by 32%. But let's give our hungry ewe some credit and assume that she can *partially* compensate by slightly increasing *both* her grazing time and her bite rate. Thus her actual dry matter intake may only be 20% less than NRC expectations, rather than 32%.

Let's follow this up. A 20% reduction means that her actual dry matter intake will only be 5.0 lb of forage, not 6.2. These 5.0 pounds will provide only 3.5 pounds of TDN and 1.0 pound of crude protein. These numbers show that our ewe still receives enough protein from the forage, *but her total daily intake does not contain enough energy, even though the forage nutritional value appears excellent when reported on a dry matter basis.*

Although our ewe is grazing happily on this forage, she won't make enough milk to support high growth in her twins. She cannot consume enough energy in that lush forage to supply her needs. This is a fancy way of saying that our winter pastures are "too washy," which is not exactly a new concept in this county.

So, aside from trying to dry out the pasture with a humongous hair-dryer, what can we do about it?

Offering grain to ewes or creep feed to lambs may not be a good option for us at this time of year. Corn certainly contains energy, but it also costs money and labor. Grain-feeding can also cause problems like crowding, mud, coccidiosis, foot disorders, and lambs trampled by rushing ewes.

But maybe we can, in a sense, dry out the pastures — by allowing the forage to get a little older before we put animals on it. At 3 inches, our fescues and ryegrasses are too washy, but at 7–8 inches, their dry matter levels have risen above 20%, which is close enough. Each bite of this older forage would include a little more nutrition, and day in and day out, that makes all the difference. And if we have to wait for the grass, we just need to feed hay for a little longer.

Or we can do nothing — which is what most folks do. The grass is green, the lambs are growing, and our labor and feed costs remain low. But because the reduced milk output comes at the beginning of lactation, the entire milk production curve is lower, so the lambs will grow slower throughout lactation. And, like every other year, the lambs will be lighter at weaning.

First Published: March 1995

Author's Note: Different sections of the continent face this problem at different times, depending on their climates. But the principle is the same — the lower dry matter in the forage must be compensated by increased intake, either in grazing time, grazing speed, or both. Incomplete compensation translates to reduced nutrient intake. The rest is just elementary . . .

Lambs & Calves & Milk, Oh My!

When I first started in the sheep business I thought that orphan lambs were called "bummer lambs" because after you spent weeks raising them and feeding them, they often died, which, in the parlance of the day, was a real "bummer." Scours, digestive problems, clostridial diseases — you name it, orphan lambs succumbed to it. But we can do better today; we have the knowledge and the tools . . .

Our basic tool for rearing orphan lambs is milk replacer, which is just a milk substitute. So let's look closely at natural milk — its composition and its consumption patterns — and observe how these factors may affect our rearing techniques.

Milk is made up of five basic components: water, fat, protein, lactose, and ash. Lactose is the well-known "milk sugar." Ash contains all the minerals.

But milk isn't just milk. Each species has its own recipe for success. Milk composition varies quite dramatically from species to species. Our familiar cow milk generally contains 87% water, which equates to 13% dry matter. This dry matter is composed of 29% fat, 27% protein, and 38% lactose. (It also contains 6% ash, but we'll ignore that during this discussion because ash variation is not relevant to our orphan lamb problems.)

Ewe milk, however, is quite different from cow milk. Ewe milk contains 19% dry matter, which consists of 41% fat, 29% protein, and 24% lactose. Therefore, compared to cow milk, sheep milk has more fat and dry matter and less lactose. In fact, cheese manufacturers would love to get their hands on sheep milk — its lower water content means higher cheese yields, and its fat has a delicious flavor.

Species variability doesn't stop there. Human milk contains 35% fat, 8% protein, and 56% lactose. Horse milk contains only 15% fat, but it has 25%

protein and 55% lactose. Kangaroo milk contains 37% fat, 47% protein, and 10% lactose. But all these differences pale when compared to the composition of whale milk (yes, whales are mammals too). Unlike a four-legged cow, a whale can't just stand around, chew its cud, and wait for its young to finish suckling. A whale has places to go, things to see — and therefore it must transfer a lot of concentrated energy to its calf in a short period of time. Also, a whale cannot use too much internal water to manufacture milk because of the osmotic constraints of living in salt water. So whales have evolved a relatively "dry" milk containing 57% dry matter, which is composed of a whopping 73% fat, plus 21% protein and only 2% lactose. What a potential for cheese! If only we could design some practical milking equipment for whales, and while we're at it, a good headgate.

From an orphan lamb's point of view, however, the critical issue is lactose. Lactose is a type of carbohydrate called a *disaccharide* which means that it is made from two sugar units strung together by a distinctive chemical bond. Animals secrete an intestinal enzyme called *lactase*, whose sole purpose is to split that bond. (This is the same lactase enzyme that lactose intolerant people don't have.)

Only after this chemical bond is severed can those individual sugar units be absorbed across the gut wall into the blood. If for any reason there is not enough lactase in the intestine, the lactose molecules remain intact and proceed merrily down the gut until they reach the large intestine. Bacteria in the large intestine are very happy to metabolize these lactose molecules into their own bacterial products. This bacterial fermentation of lactose is usually no big deal for lambs, unless it occurs in large amounts. Then, however, it can cause scours.

Lambs are not calves. Calves come into this world equipped with large amounts of lactase, but lambs do not. This means that under typical rearing conditions, calves can easily handle the higher levels of lactose in cow's milk, even if they are fed only twice each day, but lambs can run into trouble if their digestive systems are overwhelmed with these amounts. Commercial manufacturers have recognized this principle by designing special lamb milk replacers that contain less lactose than calf milk replacers.

But we also have a second tool for raising orphan lambs: our knowledge of how lambs consume their milk. Digestive problems with lactose are not due only to the actual *amounts* of lactose, but also to the *timing* of these amounts.

Let's consider, for a moment, that you are a cell in a lamb's gut wall. You must deal with lactose whenever it comes down the pike, so to speak — which is whenever the lamb drinks milk. If you don't have a lot of lactase to secrete, you have to hope that not much lactose comes at you all at once. So in nature, how many times does a lamb normally suckle its ewe each day? The answer — based on some very exhausting research — is fifteen to twenty times! Each suckling session may not last very long, but over 24 hours a lamb will still consume a lot of milk. In nature, therefore, a lamb's pattern of milk consumption matches its low intestinal level of lactase. Its digestive tract cannot easily process huge surges of lactose. In short, a lamb's entire system of behavior and physiology is designed to handle relatively small amounts of lactose at any one time.

But now reality: How do we usually rear orphan lambs in our real-world barns? We give them TLC (a highly technical term meaning "tender loving care") and hand-feed bottles of milk replacer two, four, or six times each day. As our orphan lambs get older and bigger, we increase the amount of milk replacer at each feeding and decrease the number of daily feedings. And if the price of lamb milk replacer becomes too high, we sometimes try to use calf milk replacers to cut costs.

Lambs are not calves. But with our management, we try very hard to make our lambs into calves. Then we wonder why lambing can sometimes be the saddest of seasons.

There *is* a better way — we can use our tools. We can think like a ewe and treat orphan lambs like lambs. We can use only milk replacers designed for lambs. We can follow a self-feeding regime that allows a lamb to duplicate its normal feeding routine — so that lambs can suckle whenever they want. And we can use cool milk replacer (40–50°) to discourage engorgement.

That way, orphan lambs will have a better chance of surviving — and they won't be such a bummer.

First Published: January 1995

Author's Note: It would seem to be such a simple adjustment — don't feed orphan lambs like calves. I know ranchers who raise 300 or more orphan lambs just fine, but they scrupulously follow this advice.

Growing Once . . .
Growing Twice . . .

Young lambs eat at only one speed — fast. They are like nutritional speedsters, always barreling along in high gear. Ewes have lots of milk, so isn't this a match made in heaven? At first, certainly, but what happens after the initial surge of milk is over, at around 60 days?

Let's look at this 60-day situation carefully — the requirements of a voracious, growthy lamb *versus* the contribution of its mother's milk. Should be interesting.

Our sample lamb entered this world in April as a single 12-pound ram lamb. Its pedigree is impeccable: a composite hybrid containing seven different breeds as its dam and one humongous fast-growing blackface breed as its sire. For reasons beyond its control, our lamb soon became a wether. Except for that minor setback, it grew well and, because of its good rearing and family lines, weighed a solid 55 pounds at 60 days of age, which equates to a healthy average daily gain (ADG) of 0.72 pounds. Although its mother received no grain after lambing, she grazed high-quality pasture throughout the spring.

At 60 days of age, what does our 55-pound lamb need nutritionally to maintain this good growth rate? I go to my copy of the *SID Sheep Production Handbook* and look at the reference table of nutrient requirements. This table only has categories for lambs of 44 and 66 pounds, but I can do a little arithmetic, so I'll split the difference. The numbers are pretty straightforward: an ADG of 0.7 pounds requires a daily consumption of 2.2 pounds of TDN (energy), 0.47 pounds of crude protein, 6.8 grams of calcium, and 3.2 grams of phosphorus. (Plus, of course, all the other trace minerals, but we'll ignore them in this example because the lamb will always have access to a good trace mineral mix.) By the way, if given a chance, our ravenous lamb can consume dry matter at a rate of nearly 5.3% of its body weight (= 2.9 lb).

Now, let's look at the other side of the milk equation — let's look at the ewe. Frankly, ewe's milk is an incredibly good feedstuff. We can find its nutritional composition listed in many reference tables. Generally ewe's milk contains 19% dry matter, which in turn contains 161% TDN, 24.7% crude protein, 1.05% calcium, and 0.79% phosphorus. (Yes, TDN *can* have values greater than 100%. Remember that fat contains 225% TDN, and milk contains lots of fat.)

The critical question is *how much milk does our ewe produce at 60 days*? We know from many research trials that a ewe's lactation curve peaks at 20–30 days after lambing and then slopes downwards gently (or quickly, depending on the ewe). Let's get specific. I combed quite a few scientific sources, which obligingly gave numbers that varied all over the map — probably just like the maddening variation we see in ewes on pasture. But I'll settle on the rough estimate of 4.4 pounds (2,000 g) of fluid milk at 60 days. This may be on the high side, but let's give our ewe the benefit of the doubt.

This 4.4 pounds of liquid milk translates to 0.84 pounds of dry matter (= 4.4 x 0.19). Thus, at 60 days post-lambing, our ewe pumps out each day 1.35 pounds of TDN, 0.21 pounds of crude protein, 4.0 g of calcium, and 3.0 g of phosphorus (for a homework exercise, you can do the calculations).

Now let's assume that our lamb can drink all the milk that its mom produces. I think this is a fairly safe assumption, since 0.84 pounds of milk dry matter is well below the lamb's daily limit of 2.9 pounds. With that assumption under our belt, we can now estimate how well the milk fulfills our lamb's nutritional requirements for good growth.

In short, the answer to that question is: not very well. Here are the numbers: the 1.35 lb of TDN from the milk equals only 61% of the lamb's energy needs (= 1.35 ÷ 2.2). Similarly, the ewe's milk provides only 45% of the lamb's requirement for crude protein, 59% for calcium, and 94% for phosphorus. At least the phosphorus is close, but that may even be a double-edged sword, because the calcium-phosphorus ratio in the milk is only a marginal 1.3:1.

At 60 days of age, our lamb is a full-fledged, card-carrying ruminant. It chews its cud, digests fiber, and grazes forage just like the big boys. If the milk supply is deficient, our lamb can and will eat other things — creep feed, pasture, etc. But most producers don't provide creep feed on pasture, so our lamb usually doesn't have that nutritional option. And on pasture, when our lamb accompanies its ewe — after being raised properly by that ewe — it doesn't wander too far from her, which means that they both hang out in the same area, which means that they will compete for the same patches of forage. As the summer deepens, that forage will become less nutritious and less plentiful. And the lamb, which continues to depend on the ewe for much of its nutrition, simply will not get enough groceries to support an ADG of 0.7 pounds. Instead, it may gain only 0.5 pounds per day, or less. This situation only gets worse as our lamb gets larger and the ewe's milk production continues to drop. The genes that would permit its high rate of growth haven't disappeared; they've just been masked by the lack of nutrients.

Our example has been a single lamb. If our lamb had been a twin, also with good genetics, its plight would be even worse. Although a twin lamb would still have the same good genetics for growth, and a ewe raising twins *does* produce more milk than a ewe raising a single lamb, this amount *isn't* twice as much as for a single. Thus each twin lamb receives *less* milk daily than a single lamb, especially by 60 days. Well, we've always heard that twin lambs don't grow as fast as singles.

There is an alternative: the ewe could be weaned and the lamb put onto better feed — like on high-quality pasture or in a drylot. Pasture is really an attractive option because forages are cheaper than grain. And good, well-managed pasture can have the energy value of oats with nearly twice its protein.

Well, it's certainly something to ruminate on . . .

First Published: August 1997

Author's Note: These nutrient requirements were derived from the 1985 edition of the NRC *Nutrient Requirements of Sheep*. Although the new 2007 edition may contain slightly different number, the basic pattern of intake and need is the same, and so is the result.

When Do We Wean Turkeys?

Lambs are more adaptable than you think. A few years ago I spent a couple of summers conducting a research trial in which I weaned lambs onto alfalfa pastures at less than 30 days of age. The lambs did fine; they grew well and hardly noticed the change. It was the people who gave me a hard time — in spite of the facts, they all said that it couldn't be done.

Let's ruminate about rearing young ruminants.

Cattlemen routinely separate calves from their cows at sale time, at about seven months of age. When cattlemen talk about early weaning, they're really discussing weaning their beef calves at about five months (150 days). In contrast, consider a typical dairy calf. A dairyman separates the calf from its mom soon after birth, places the calf in an indoor stall or outdoor hutch, feeds it milk for only six weeks (42 days), and then weans it onto a ration of grain and a little bit of forage. The dairy industry considers this management as relatively standard. Professional calf rearers — people who make their living by raising young dairy calves and selling weaned animals to dairymen or cattle feeders — are always looking for ways to reduce their feed costs. *They* routinely wean their calves at 31–35 days. Researchers push the system even further — they run all sorts of experiments on calves weaned at 21–28 days. I've personally weaned dairy calves onto solid diets at 26 days, even calves reared in outdoor hutches during a Wisconsin winter.

The sheep industry has parallel situations. Range operators really don't wean their lambs at all. Like their counterparts in the beef industry, range operators just separate lambs from ewes when they sell the lambs — usually when the bands come down from the mountains when the lambs are at least five months or older (150+ days). "Early weaning" to a range operator can mean 120 days of age. Sheep producers who manage intensive operations, however, think nothing of weaning their lambs at 60–90 days. In fact, modern sheep genetic evaluation programs like NSIP (*National Sheep Improvement Program*) assume that lamb growth after 60 days shows the lamb's own genetics for gain, rather than the ewe's milking ability. (Ewe milk production peaks at about 30 days after lambing and tails off after 60 days.)

But the buck doesn't stop there. Sheep producers on accelerated lambing programs, like the STAR System, must wean lambs between 45–60 days. Orphan lambs are pushed even harder. A 1980 USDA publication recommended that artificially-reared lambs should be weaned at 28–35 days. Today, many advanced producers routinely save milk replacer and veterinary expenses by weaning orphan lambs at less than 21 days. And of course, research trials have squeezed the system even more — researchers have successfully weaned orphan lambs at 14 days or even less.

There's a pattern here. Lambs can be pushed, and we can benefit financially if this pushing fits our system.

And the system doesn't always have to be based on grain. If we are running a pasture-based system, we still have lots of options.

Consider a lush legume pasture, like young alfalfa or clover. Unlike blades of grass, the leaves of young legume plants don't contain much fiber. In fact, those legume leaves are very palatable and high in protein, calcium, and digestible energy. If you stretch your imagination, you can even classify young legume leaves as a type of creep feed, albeit in a slightly different form than the usual grain mix. And if you put very young lambs or calves onto a pasture that contains lots of young legume leaves, what do you think they will eat?

A little background physiology here: When a milk-fed lamb consumes milk, the liquid does *not* go into the rumen; instead it travels into the true stomach (also known in technical circles as the *abomasum*). Young ruminants are equipped with a flap of tissue called the *esophageal groove* which acts as a bypass valve to shunt liquids away from the rumen toward the abomasum. This bypass valve does not affect solid food — it only affects liquids. So when a young ruminant consumes any solid food, that material travels directly into the rumen, just like in an adult ruminant. Although this rumen is not yet fully functional in very young calves and lambs, nonetheless, anything deposited in it *will* begin to ferment. This fermentation produces compounds such as acetate, propionate, and butyrate. These compounds, especially butyrate, stimulate the natural process of rumen development. The entire process is therefore a positive feedback system: as more solid feed ferments in the rumen, the faster the rumen develops its physiology and anatomy.

Weaning healthy lambs and calves has a universal effect: they get hungry. And if they are offered a diet that is nutritious and palatable, they eat. Young legume pasture admirably fits the bill.

And that's what happened to my lambs when I weaned them onto a good pasture at only 28 days of age. They ate, and they grew. I dried off the ewes mercilessly — keeping them on clean, dry ground and watching them carefully for the signs of mastitis, which never appeared.

There is an old belief that lambs and calves must have good rumen development *before* they can be weaned. That's not true. It's the other way around — the weaning process itself speeds rumen development.

Early weaning is not for everyone, of course. It is a delicate management system that requires good judgment and meticulous observation of details. It *can* give some operations a financial edge, but old notions must be discarded first. Success means providing a high-quality diet and managing the animals so that they want to consume that diet. In other words, *we should feed very young lambs and calves as if they don't have rumens.*

But if you stop and think for a moment, we are already doing all this for other species of livestock. After all, when do we wean turkeys?

First Published: June 1995

Author's Note: I still read articles in research journals and popular magazines about early weaning in which the authors focus on improving rumen development. I know hundreds of farmers and ranchers who are ahead of that game. When will we ever wean those authors from their outdated thinking?

Points of Weaning

My office wall is lined with dusty shelves of old electronic messages that sometimes yield pleasant surprises. Rummaging around these artifacts, I unearthed an email that I had once written about young lambs and weaning. I blew off the extra electrons and read it carefully. Hmmm, it was pretty good — a list of critical nutritional principles that apply to *all* lambs, orphans *and* ewe-reared. I thought to myself, "Self, maybe someone ought to publish this." Then I realized that indeed *I* could publish it, here, in this article. So this month, I'll share these principles with you — dependable physiological notes that you can use to make weaning choices about your lambs. My emphasis is on early-weaning because it's economically sound and it works.

But FYI, I changed the original message a bit: I fleshed out some of the terse email sentences and also eliminated the nerdish acronyms that clutter up so many emails (OTOH, 4U). I retained the bullet points because they are easy to read. Enjoy.

- You can wean lambs earlier than you think. Skilled shepherds routinely wean orphan lambs at three weeks, but at that age, weaning has to be done *right*. Ewe-reared lambs can be routinely weaned at 60 days, as long as the lambs are healthy and the mothers are managed properly after weaning to reduce the risks of mastitis. Any good sheep book will discuss the show-and-tell details about these procedures. My next paragraphs focus on the physiology of why they work.
- Weaning is really the transition from one diet to another: from a high-fat, high-protein, high-sugar liquid diet to a low-fat, lower-protein solid diet. Lambs will do this naturally over time, in their time, but unfortunately, a lamb's natural internal clock does not usually coincide with the profitable economics of a sheep operation. Hence my focus on pushing the system and weaning early.
- From a gastrointestinal perspective, young lambs resemble dogs and cats more closely than adult sheep. The rumen of young lambs is tiny

and relatively unformed, and little or no fermentation occurs in it. Young lambs are essentially *monogastric* animals, like dogs and humans, and they need high-quality feed that can be digested without relying on fermentation. Like milk.

- Milk (and milk replacer) never enters the rumen. When a lamb sucks a nipple, a small flap of tissue (called the *esophageal groove*) — like a bypass valve — closes off entry into the rumen. Any milk traveling down the gullet bypasses the rumen and goes directly into the true stomach.
- Although milk is a great feedstuff, ewe-reared lambs will naturally begin consuming significant amounts of dry feed at around 3–4 weeks. Even if you see younger lambs "chew" something (usually over and over and over, as young lambs like to do), that amount is tiny and biologically insignificant. But after three weeks of age, the amounts get larger. Especially if the feed is a ground, high-protein, tasty grain mix. Orphan lambs will consume dry feed earlier than four weeks if they are weaned by then.
- When lambs consume dry feed, this material is not rerouted by the esophageal groove. It goes directly into the rumen where it is exposed to microbial fermentation.
- Here's a positive feedback system: the consumption of dry feed *stimulates* rumen development. Dry feed is fermented by bacteria and protozoa in the rumen, even in a small, immature rumen. This fermentation results in products (primarily butyrate) that stimulate additional rumen growth and development. As more feed enters the rumen, more stimulation occurs.
- But — here's an intriguing point — *at first, the rumen is actually not very important*. A lamb does *not* need a functioning rumen to consume and digest dry feed. High-quality feeds are easily digested in the small intestine, and those digested nutrients will support good health and growth. If it helps, you can think of milk as a high-quality dry feed with water added. (Actually, the milk replacer powder in a bag *is* a dry feed). If you don't believe me, ask yourself — do *you* chew cud? If you don't and are reading this, you are living proof that a rumen is not necessary for the good life.
- A functioning rumen is really only necessary for a few things: (1) fiber digestion, (2) utilization of feedstuffs containing nonprotein nitrogen (like urea), (3) detoxifying certain harmful compounds before they reach the small intestine, and (4) production of B-vitamins. If lambs don't need to accomplish the first three things to survive or grow, and if their high-quality feed contains ready-made B-vitamins, the lambs will thrive.

- Historically, in an effort to facilitate early weaning, many scientists have focused their research on accelerating rumen development. This is an interesting case of misplaced emphasis, because a lamb doesn't need a functioning rumen for successful weaning. The main purpose of weaning is to shift the diet from milk to solid feed. Rumens can develop quite well on their own *after* early weaning.
- Lambs on high-quality vegetative legume pastures will consume leaves and young shoots. The nutritional value of this forage can equal or surpass most grain-based creep feeds. These legume plants will definitely support the healthy growth of young lambs after four weeks of age.
- Milk production peaks at 3–4 weeks after lambing. By 60–70 days, milk production is relatively low compared to the nutritional needs of a growing lamb. (Udder size is *not* a good indicator of milk production, ask any dairy producer.) After 60 days, ewe milk will provide less than 50% of a single lamb's nutritional needs. This percentage is lower for twins, and even lower for triplets.
- When ewe-reared lambs get older — after 50–60 days — they compete directly with their moms for feed, especially when a flock grazes high quality pasture. Think about this for a moment. Ewes *never* defer good pasture to their lambs. Even the best mom will *not* step back to allow her lambs first choice of clover instead of her. Since ewes are bigger than lambs, ewes consume more forage than lambs. If good pasture is limited, or if you are juggling pastures in an intensive grazing system with a flock of ewes and their unweaned lambs, be aware that those ewes will consume a high percentage of your best-quality forage.
- In late lactation (after 60 days), a ewe-lamb flock is really composed of *two* subflocks with wildly different nutritional requirements: (1) young growing lambs, and (2) ewes in late lactation. And if milk production is very low, those ewes are effectively in maintenance mode. If the lambs and their ewes are grazing together, how can a shepherd meet the nutritional needs of both groups at the same time, in the same paddock? Either one group will be overfed or the other group will be underfed.
- When to wean? Decisions, decisions. But *not* making a decision is still making a decision, because if you make no changes, the ewe-reared lambs will *continue* to compete with their ewes for feed, and the orphan lambs will *continue* to drink expensive milk replacer. Physiology, nutrition, pastures, timing — it's *all* economics.

First Published: February 2004
Author's Note: The principles of weaning young ruminants from milk to dry feed are universal. Applying them to orphan lambs or to ewe-reared lambs are just details. Of course these details have to be monitored carefully, but then again, that's part of the art in animal husbandry.

The Business of Early Weaning

At any meeting, the speaker who brings up the topic of weaning lambs early at 60 days will elicit strong opinions from folks in the audience — about rumen development, loss of bloom, type of grain, risk of mastitis, etc. But early-weaning is a management technique with many layers, and all the associated hullabaloo often obscures some of its more compelling benefits. I'd like to talk about two of these benefits — grazing and parasite control.

Grazing: Early weaning can help with good pasture management. Weaning, by definition, creates two different flocks with wildly different nutritional requirements — dry ewes and weaned lambs. Dry ewes need only maintenance levels of nutrition while growing lambs need high levels of protein and energy. We can choose to manage these groups as two separate units or as a single mob. Weaning separates them into two units. If we don't wean, we are really choosing to manage them as a single mob.

Keeping all the animals together in a single mob — i.e., not weaning — means that some sheep will be underfed and some sheep will be overfed *at the same time.* As forages mature in the pasture and lose their nutritional value, the lambs who need more nutrients for growth are underfed, but the adult ewes, who only need a low level of nutrition, are overfed and will simply gain weight.

But weaning creates a separate mob of dry ewes, and this provides us with all sorts of good management options: (1) we can feed them old, poor quality hay, (2) we could put them on a paddock that we *want* to drive into the ground, like a weedy area infested with thistle, blackberry, and poison oak, or (3) we can use them to renovate a paddock by broadcasting seed onto the ground prior to grazing, stocking densely, and letting the hooves pound in the seed. Our choice depends on our management system and also our ability to use temporary electric fence. A flock of dry ewes can be a precision instrument for managing pastures.

Fast-growing lambs, on the other hand, need high-quality feed. Lambs born in winter or early spring are at least 45 days old when spring pastures are first ready for grazing. These lambs already have well-developed rumens; they can easily consume good quality forage and use it efficiently. But when the unweaned lambs continue to hang around their ewes, *all* the animals — lambs *and* ewes — compete for the same forage plants. This becomes critical when pastures begin to mature or when the amount of forage is limited. In all the years I've worked with sheep, I've *never* seen a ewe — even a great mother ewe — step *back* and give her lambs first choice on a patch of clover. But once the lambs are weaned, we can move them into a new paddock without worrying about competition from the ewes.

Early weaning, therefore, allows us to provide high-quality feed to the lambs and reduces the expense of purchased grain. Good pasture management is often a matter of timing: by reducing the grazing pressure at critical times, we encourage higher production of good quality forage, which can cut our costs of making, storing, or purchasing winter feed.

The second benefit of early weaning hits the checkbook even more directly: how much does it cost to worm a flock? What if we can save more than 50% of that cost without compromising the health of the lambs?

Parasite control: As every shepherd knows, sheep on green pastures usually become infested with intestinal parasites (primarily roundworms, also called *nematodes*). Adult ewes generally resist the effects of these worms because they have developed some immunity, but lambs with their immature immune system can be devastated by them. Periodic doses of deworming medicine reduce that worm load. Veterinarians in most areas of the country currently recommend treating sheep every three weeks (which is the length of the life cycle of some of the major nematode species), particularly early in the season when the lambs are younger and the lush pasture growth is ripe for worm larvae.

An accepted principle of parasite control is to dose *all* the animals at the same time; we can't just treat the "wormy-looking" ones. This means that in a flock containing unweaned lambs, we must worm the lambs *and* all the ewes. Otherwise, the ewes would just continue to release millions of viable worm eggs back onto the pasture.

Here's the rub: *worming medicine is given by weight*. Heavier animals receive larger doses than lighter animals, and worm medicine is not cheap. Since ewes weigh more than lambs, it costs more to dose ewes than lambs. This additional cost is essentially an overhead expense.

For example, let's say that we are pasturing 100 ewes with their 150 unweaned lambs (= 150% lamb crop before dogs). The ewes average 160 pounds and the lambs average 50 pounds. Our flock, therefore, consists of 16,000 pounds of ewes (= 100 x 160) plus 7,500 pounds of lambs (= 150 x 50), giving a *total flock weight of 23,500 pounds — 68% of that weight in ewes and 32% in lambs.* In an unweaned flock, therefore, 68% of our worm medicine goes to the ewes, which represents a steep overhead expense.

Let's speak in dollars: I recently purchased a 960 ml bottle of Ivomec® (Ivermectin) for $55.00, which means that 1 ml of Ivomec® costs $0.057 (5.7¢). The label instructions are explicit: orally drench 3.0 ml of Ivermectin for each 26 pounds of sheep (= 0.115 ml of wormer per pound). Therefore, the dose for a 160-pound ewe would be 19 ml, costing $1.05 (= 160 x 0.115 x $0.057). This adds up to $105.00 for all our 100 ewes. In contrast, a 50-pound lamb would receive only 6 ml, costing $0.33 (= 50 x 0.115 x $0.057), which adds up to $49.50 for 150 lambs. Each time we worm this unweaned flock of lambs and their ewes, we spend a total of $154.50 for drugs, which includes more than $100 just to cover the overhead of these ewes.

But dry ewes really don't need worming medicine until late autumn or winter (local veterinarians can help with the details). In general, dry ewes won't pick up many parasites if they are grazed on parking lots or other poor-quality pastures. For our sample flock, we can save $100 at each worming by simply moving the ewes away from the lambs — i.e., weaning.

And these calculations don't include the time savings gained by having fewer animals to handle and drench.

Efficient businesses try to find ways to cut overhead costs without hurting their production. Early weaning can be a sound business strategy for reducing that overhead.

First Published: May 1997

Author's Note: This is a good example of how things change. Since I wrote this article, the concept of FAMACHA has become widespread for the control of *Haemonchus contortus*, which is the most serious intestinal parasite for sheep. The FAMACHA system requires treatment of only the most infected animals, as determined by an anemia score (eyelid color), not the entire flock. But the advantage of early weaning still applies — by sending the ewes somewhere else, you can concentrate your time and money on controlling parasites in the vulnerable lambs rather than in the entire mixed flock of ewes and lambs. The other thing that changes is price. The current cost of Ivomec? I don't even want to list it. But at least the principles don't change.

Protein Arithmetic

How much protein are you feeding your animals? How do you calculate it?

This is not an academic question. Protein is crucial. Protein helps determine growth rate, milk production, or if that cow or ewe will even *think* about conceiving this month. I've seen ranchers sit up all night discussing protein — ahem — well, maybe helped out a little by a hand or two of cards. In any case, some knowledge about protein will help us decide what is going on with our livestock . . . and with our bank account.

Let's take a practical example from here in Douglas County, Oregon: a 5-month old weaned lamb on a non-irrigated pasture in early June, already weighing 100 pounds. Our lamb is thoughtfully attempting to save its owner from financial embarrassment by trying to gain enough weight to finish at 115 pounds. (This may be a rather shortsighted goal from the lamb's point of view, but it will suffice for our example.) So, how can we determine if things are going okay, protein-wise??

First, we establish what the lamb is eating and how much. Second, we determine the protein level of each of the dietary ingredients. How? One of three ways: (1) we actually test these ingredients for protein, or (2) we look up their protein values in reference tables, or (3) we make shrewd estimates ("guesses"). Finally, we add up these numbers and compare the total protein intake against the lamb's needs (requirements). Simple . . . on paper.

Let's return to our thoughtful lamb. On a June pasture, it probably consumes 3% of its body weight in dry matter.

An important nutritional aside: dry matter is the feedstuff without its water. It's as if one could magically remove the water molecules from the feedstuff and leave everything else. Actually we can do this arithmetically by dividing the *as fed* value by the dry matter percentage (expressed as a coefficient). Water contains no energy or protein, but its weight masks the true nutritional value of the feed. Think of fresh cow's milk, which is certainly a high-quality feed, yet it only lists at 3.3% protein because it contains so much water. Even stored hay has 10–15% water in it. So nutritionists ignore the water and talk about feeds on a *dry matter basis*. This convention simplifies things and allows us to compare feed values and animal requirements on an equal footing. Trust me.

Reference tables usually indicate that our lamb could potentially consume more than 3% of its body weight. That's true — for diets primarily composed of corn and supplements, or of very young March pasture that contains little fiber. In contrast, our lamb is grazing an early June pasture, which contains high levels of fiber, which reduces intake. Realistically, if a lamb ate more than 3% of its body weight of this pasture, it would explode.

Let's be generous and assume that a typical, non-irrigated Douglas County pasture in early June contains about 12% crude protein. Or if there is lots of clover, maybe 14% (on a dry matter basis, remember). This forage is already fairly mature, with many plants headed out. Yes, I know that some pastures in this county typically dry up after only two hours of green growth, and those paddocks would even be worse than our hypothetical example. But nonetheless, let's assume that our pasture received some fertilizer in the recent past and also received good grazing management. Of course, our lamb is also consuming a small amount trace mineralized salt (isn't it?), but we can ignore this because it contains no protein.

Now the final key steps: If our 100-pound lamb eats 3% of its body weight in forage dry matter, it will consume 3.0 pounds of forage dry matter (= 3% of 100). If the forage contains 12% protein, this means that our lamb would consume 0.36 pounds of protein (= 0.12 x 3.0). Similarly, if the forage contains 14% protein, our lamb would consume 0.42 pounds of protein (= 0.14 x 3.0).

Now we must compare these intake amounts against the lamb's protein requirements for growth, which we can find in any handy reference book. We can't guess at these values; we must look them up. I usually use the Nutrition Chapter in the *SID Sheep Production Handbook*. In any case, assuming that nothing else is lacking in the diet, 0.36 pounds of protein will support a growth rate of 0.4 pounds per day. The higher clover pasture, with its 0.42 pounds of protein consumption, will support a growth rate of 0.6 pounds per day.

We immediately see the practical value of clover in a pasture at this time of year — the extra protein can really help the gain. Even these simple numbers demonstrate a 50% increase of gain.

But we may still have a problem. In this situation, our lamb must reach 115 pounds to be market-ready. Since it currently weighs only 100 pounds, it therefore must *gain* 15 more pounds. At a growth rate of 0.4 pounds per day, it will still need 37 days to add those 15 pounds (= 15 ÷ 0.4), which will put us into the middle of July. And during these 37 days, our non-irrigated Douglas County pastures become more and more mature, kind of like standing straw. Without any supplementation, what will then happen to our lamb's growth rate?

So it shouldn't surprise us when lambs stall during late May and June. No matter how much they want to help their owners.

First Published: June 1994
Author's Note: So how much protein are you feeding your animals? Isn't this a common question? The growth lesson is a good one. In this article I used the *SID Sheep Production Handbook* with reference tables derived from the 1985 Sheep NRC document. If we used the newer 2007 Small Ruminant NRC document, we would find that the protein requirements would be slightly different, as well as expressed in a more obtuse way, but this is a story for a different time.

To Grain or Not To Grain

Recently, I was scheduled to give a presentation to shepherds on the topic of getting their lambs to market faster by supplementing grain on pasture. I'm sure that you've already heard the routine: supplement grain to pasture lambs, increase their daily gains, reduce time to market, improve the finish on those lambs, receive higher prices from the packer, and increase profits.

This seemed straightforward enough . . . at first. But then I started thinking about it carefully: feeding grain to pasture lambs *may indeed* be a clear-cut strategy to increase profits, but then again, *it may not*. The more I thought about it, the more I realized that it was anything but clear-cut. Here is why:

Let's say that you're grazing 30 or 100 or 800 weaned lambs on pasture. Prior to weaning, you may have offered them creep feed, or not, and you may have raised them on pasture with their ewes, or not — depending on when you lambed and how you fed your lactating ewes. In any case, at this point in the year, you've already weaned those lambs, and now they're on pasture. We are now in the middle of the growing season, warm enough for forage growth, with our lambs gamboling and frolicking in the grass. (I'll use the word "grass" as a generic term for any forage-based grazing situation, including alfalfa, brassicas, grass, or overgrown weeds.)

Here is the potential scenario: You are thinking of offering, say, a half-pound of a 16% grain supplement to these lambs on pasture (that's per lamb — what were *you* thinking?). For 30 lambs, that's 15 pounds of grain per day; for 800 lambs, it's 400 pounds of grain. Your pasture is reasonably good quality, vegetative, with 14–16% crude protein and a TDN value of at least 60%. Your records show that in previous years, your pasture lambs averaged 0.35 lb/day on this grass through the growing season. Not bad, but not earth-shaking. Your records also indicate that at the end of each growing season, you typically sold 50% as finished lambs and then you fed out the rest near the barn until they were sold during the winter. Now you are considering an alternative strategy of supplementing grain on pasture, which promises a higher percentage of finished lambs earlier, with the hope that your buyers will eagerly pay top dollar.

Let's review the steps in this grain-feeding program: (1) supplement grain, (2) increase daily gain, (3) reduce market age, (4) improve finish, (5) receive higher prices, (6) increase profits. Now let's examine each one.

Supplement grain. Providing grain to lambs on good-quality pasture may not be as simple as it appears, especially if you have lots of animals. Firstly, from a nutritional perspective, grain doesn't simply add energy into the ration. The half-pound of grain will *replace* some grass intake, rather than simply adding to it. Also, rumen fermentation of the grain starch can reduce the rumen pH, at least temporarily, which can reduce the digestibility of the rest of the forage in the diet. Textbooks call this an *associative effect*, but whatever they call it, it reduces the financial value of the grain.

Secondly, from a practical perspective, you feed grain by either (a) bringing the grain to the lambs, or (b) bringing the lambs to the grain. For option (a), someone has to do this, every day. That may work nicely for a person with only a few lambs and lots of extra time. But if you have 800 lambs and are trying to juggle a zillion other tasks in your operation, time is something you *don't* have.

If you still like option (a) and want to bring grain to the lambs more efficiently, you could use a self-feeder and load it periodically with grain. But this technique also entails risks — some lambs may eat too much and suffer from acidosis, or you might see your best, most ravenous lambs die from enterotoxemia because the vaccine wasn't 100% effective (for all sorts of reasons we won't go into here). Therefore, you would need to formulate a supplement which contains something to reduce acidosis, like sodium bicarbonate, or . . . well, things can get complex. It's not that these problems will *always* occur, but self-feeding grain introduces risks which were not previously part of your world.

If, on the other hand, you choose option (b), someone has to drive the lambs to somewhere each day for their grain. If you manage your pastures with intensive grazing, then you will need to adjust fences and/or build lanes to allow easy lamb movement to the barn. If you set-stock those pastures, the lambs will tend to hang out near the grain-feeding area, which effectively transfers soil fertility from the rest of the field to that area. (I'll leave it to your imagination about how that occurs.) All these issues translate into costs.

Increase daily gain and reduce market age. Ah yes, these are the main consequences of feeding grain. But they are also a bit problematic, with lots of variability: sometimes the lambs gain okay, sometimes they show only a little extra gain, and sometimes they actually go backwards. The worst occurs if they just hang out near the feeder and refuse to walk in the hot sun for grazing. Also, their nutritional level may suffer from those associative effects mentioned earlier. And this translates into *poorer* daily gains than expected, which translates into *longer* growth periods and *higher* costs.

Reducing market age sounds good, especially if your market actually rewards you for younger lambs. If lambs are gone from the farm by early September, you certainly save daily overhead costs. Also, those sold lambs aren't around to die from predators or parasites in September or October. But marketing comes with a serious caveat — there are no price guarantees. Prices in early markets vary considerably, depending on the year and lots of other factors beyond any shepherd's control. Although theoretically, you can arrange a forward contract to solve this problem, that's usually easier said than done.

Improve finish. It's true that supplementary grain will tend to improve carcass quality, especially by adding a layer of subcutaneous fat. Within reason, this makes for a highly acceptable carcass that stores well in the cooler. Extra fat, though, is expensive. To avoid too much fat, you'll have to periodically draft off the finished animals, which requires time and a trained hand.

Receive higher prices. Well, maybe. Will packers automatically pay a higher price for slightly younger, pasture-reared lambs that have received some grain? Depends on the time of year and lots of other factors beyond the scope of this article. Grain hides a lot of mistakes, and grain can reduce the amount of variation between carcasses. That's good. But increased costs carry increased risks, and since you *must* consistently obtain a better price to cover those higher grain costs, it's even more important to cultivate a good relationship with the packer. This can be another challenge

Increase profits. Here's the biggie — the bottom line. Whatever management strategies you use, whatever clever assemblage of rations and equipment you design, *your profits are always calculated in the same way: revenue minus expenses.* Feeding grain increases your expenses — the cost of the grain, the equipment and manpower and time to feed it, veterinary costs associated with feeding it (acidosis, enterotoxemia), costs of finding alternative markets, interest on funds for buying enough grain to put into storage, and more. So . . . do the increased prices received for your lambs cover those costs and leave enough profit to reward your extra work and risks?

You know — *before* you rush out and buy a single bushel of grain for those lambs, all these factors can be listed, studied, calculated, and modeled from your kitchen table. A yellow pad, a calculator, a sharp pencil (or a spreadsheet for the computer-minded) — anyone can do this. They should do this.

Otherwise, feeding grain to pasture lambs could end up as an experience best summarized as "it seemed like a good idea at the time."

First Published: November 2006
Author's Note: Lots of workshops give the impression that high-energy supplements on pasture will push lambs faster and therefore increase profits. It's not always so.

Let Them Eat Cake

How can we get through this dry year?

Western Oregon — famous for its rain, grass, and tall trees — is experiencing a drought. Normal annual rainfall here is 35 inches, mostly falling in the winter and spring, with the summers always bone dry. People still talk about the early nineties when the rainfall was a record low of only 23 inches. This year the rainfall will be lower than that, much lower. Only 17 inches of rain have fallen since October 1 when the water season began, with no realistic prospects of more until autumn. Pastures are now drying up; yields of first-cut hay were down nearly 50%; and the summer hay crops might not materialize because irrigation water may be cut off due to low river levels and lack of snowpack. Oregon is not alone: hay yields throughout the Pacific Northwest are low. Where will we find enough feed?

The Extension Service and other agencies are conducting drought workshops. Ranchers attend and listen to lots of options: cull animals, wean early, shift dates for lambing or calving to reduce the nutritional demands of gestation, save hay by feeding grain (the classic rule of a thumb — two pounds of hay can be replaced by one pound of grain), rent land elsewhere, whatever.

But we may be overlooking one resource much closer to home. The hills and pastures in Douglas County are turning gold with dry grass. If you look closely, some of these hillsides still contain quite a bit of mature standing forage; maybe as much as 2,000 pounds per acre. Can we use this forage for livestock feed? Well, if we could force animals to graze it down to 300 pounds of residual (we're not concerned about summer regrowth because these plants are already dry and dormant), each hillside acre would provide us with 1,700 pounds of available feed. One acre could feed 60 ewes or 10 cows for nearly a week, assuming a dry matter intake (DMI) of 3.0% body weight for 150-pound ewes, or 2.2% for 1200-pound dry cows. (Good calculation exercise: $150 \times 0.03 = 4.5$ lb/day/ewe $\times 60 = 270$ lb/day/flock. Then $1700 \div 270 = 6.3$ days.)

But this hillside forage is quite mature, with high levels of fiber and *very* low levels of protein. We are all familiar with poor-quality grass hay — mostly stems, some leaves, some weeds, 6–7% protein. The standing forage on the hillsides is considerably worse — maybe only 3–5% protein, with a TDN value of less than 50%. Sure, there may be a few residual leaves, but once those are gone, the remaining feed would make an old cow want to consider retirement.

Such low protein levels cause two problems with animals: (1) these levels are lower than the protein requirements of mature ewes or cows, and (2) these levels are also below the nutrient requirements of the rumen microbes.

Actually, problem #2 may be the more serious issue, with greater economic impact. If we can alleviate this problem, even slightly, our animals and finances may endure this drought better.

When we feed a sheep or a cow, we are actually feeding two different systems — *a system within a system* — the large, four-legged system that chews its cud, runs through gates, jumps over fences, and also the complex microscopic system of bacteria and protozoa that inhabits the rumen. Both systems must receive nutrients to function well, but the *nutrient requirements* of these two systems are not identical. Dry ewes and cows can indeed survive on a 5% protein diet, although not happily or productively, but the rumen microbes really need at least 6–7% protein in the feed to thrive. Without sufficient protein entering the rumen, some species of bacteria don't function at all, particularly those that digest fiber.

Remember, the primary purpose of rumen microbes is fiber digestion. Livestock can't digest fiber directly because they don't secrete the appropriate enzymes, so they rely on rumen bacteria to do it for them. The rumen bacteria convert cellulose, hemicellulose, and other types of plant fiber into compounds that our livestock can then use as nutrients. A nice symbiotic relationship. And as these bacteria digest fiber, the fiber "disappears" — that is, it effectively "exits" the rumen, making room for additional fiber.

But when rumen bacteria don't get enough protein in *their* diet — i.e., protein entering the rumen in the consumed feeds — the bacteria don't thrive. Their populations decline, which reduces the rate of fiber digestion. When fiber remains in the rumen longer, there is less room for additional forage, and the animal's feedback system signals that it is "full." The result? An impressive reduction of feed intake. And since the forage is already low in protein, the reduced DMI makes a bad situation worse, because a low intake of a low-protein forage provides even fewer nutrients than a higher intake of that same low-protein forage. Animals lose body condition faster, and their production drops even more.

With the hillside forage at 3–5% protein, rumen microbes would indeed suffer from lack of protein. Our earlier assumptions about DMI would be too high. Instead of ewes consuming forage at 3.0% of their bodyweight or cows at 2.2%, we could expect a decline of DMI by 20% or more because of the poorer fiber digestion. Ironically, this would mean that the forage will last longer, but at what cost? The protein deficiency of our animals would be intensified, causing additional loss of body condition and severely reduced reproductive performance, etc. — problems that linger for months or, in a cow-calf operation, possibly for years.

But we *do* have a strategy for dealing with this problem. *Feed the bugs.* Give the microbes a little extra protein, or more precisely, some additional amino acids. They don't need extra energy — they will obtain *that* from the fiber. The microbes just need enough protein to meet *their* needs. Then their increased fermentation will take care of the rest. There is quite a lot of research behind this strategy.

I'll be specific. In situations where animals must consume low-protein forages, we can improve their nutrition by offering them a small amount of a protein supplement like soybean meal or cottonseed meal. A small amount means only 0.25–0.50 lb per day for a ewe, or 1–2 lb per day for a dry cow. Use any protein supplement that contains at least 25% natural protein that is easily digestible in the rumen. Cull peas or beans will do the trick. Sometimes even good quality alfalfa hay will suffice, but during a drought — well, lack of hay is one of the problems, isn't it?

Two types of supplements actually *don't* work well here. (1) Supplements containing *nonprotein nitrogen* (NPN), like urea, aren't very efficient in this situation. Urea doesn't contain true protein. Therefore, it can't provide the amino acids needed by the fiber-digesting bacteria. And (2) *energy supplements*, like corn or barley, which are low in protein also do not work. These grains will certainly delight the animals, but they will substitute for and even reduce forage intake, not increase it.

A small amount of protein supplementation accomplishes many things. The rumen microbes are happy — they ferment more of the fiber at faster speeds. The TDN of the forage increases. Sheep and cattle consume more of this forage, and their performance improves, or at least they come through the dry period in better condition with better reproductive performance. We've used our forages more efficiently while saving some costs of grain or hay.

There's a famous story about a queen who once said, "Let them eat cake!" Maybe she meant cottonseed cakes, which are chunks of cottonseed meal often used as pasture supplements because they won't blow away in the wind. Here's a royal twist on history: maybe that queen was just giving her herdsman clever instructions for feeding the royal animals through a drought year, and the royal press corps — who like many urban journalists were probably not familiar with livestock production — simply reported her statement out of context.

First Published: July 2001
Author's Note: We sometimes overlook this technique. We shouldn't.

SECTION 7

Practical Field Situations

Deep Freeze

Although cold stress may look different in different places, the principles of dealing with it are the same. In Minnesota and Montana, everyone knows what "cold" means. But if you are a sheep on a winter pasture in western Oregon with a rain-soaked fleece after four weeks of 35° temperatures, you are suffering from the same cold stress as your cousins in the big-freeze regions. And then someone drives by in a pickup and tosses some extra hay bales at you because they believe that the "heat of fermentation" of that hay will solve your problem. You try to shout "corn, barley, oats!" but you know about these humans. You sigh and baaa, and wish that you had accepted that nice job offer in a dry goods store in San Diego.

Livestock, being warm-blooded, all have a *thermoneutral zone* — a range of 30–40 degrees in which they don't suffer from either cold stress or heat stress. The lower boundary of this zone is called the *Lower Critical Temperature* (LCT). Livestock feel cold stress when the air temperature drops below the LCT. But this LCT is really a moving target because lots of factors influence it. For example, the LCT is easier to pinpoint in cattle than sheep because cattle don't wear a woolen outer garment. Cattle with a summer coat have an LCT of around 60° F. Cattle with a typical winter coat have an LCT of 32°. A heavier winter coat lowers their LCT to 19°. Wool, of course, is a better insulator than the hair of cattle (that's why sheep "wear wool"), but fleeces come in many forms: a dense 2-inch Merino fleece certainly provides a lot more insulation than an open six-month Suffolk fleece. As a general rule, a slick-shorn ewe has an LCT of around 50°, while a 2.5-inch fleece lowers the LCT to 28°. But if the wool is soaked with rain or blasted by a strong wind, its insulation value drops precipitously.

But I digress. We're really interested in the effects of cold stress on nutrition. Basically, cold stress affects two major nutritional issues: energy and intake.

Firstly, cold stress increases the *maintenance requirements* of an animal. Actually, this makes sense — cold sheep use more energy to maintain body temperature than sheep in their thermoneutral zone. A higher maintenance requirement, however, means that the animal must use more dietary energy just to maintain body heat, leaving less surplus energy to support growth, lactation, or pregnancy.

Interestingly, cold stress doesn't alter the nutritional requirements for protein, minerals, and vitamins. Which also makes biochemical sense, really. To generate body heat, animals metabolize carbohydrates and fats, not protein. Only animals in the last stages of starvation will destroy their own proteins to create heat. The bottom line is that cold stress increases the need for energy and doesn't alter the need for protein.

Secondly, cold stress induces animals to eat more feed — i.e., *their dry matter intake goes up.* Increased appetite is probably an evolutionary adaptation to the need for additional energy. (I can still hear my mom saying, "If you want to have more energy, eat more of your supper!")

Increasing feed intake, however, also increases the rate of passage of that feed through the digestive tract. This means that feed passes through the rumen faster, which reduces the time for microbial fermentation, especially of fiber, which reduces the digestibility of that fiber.

Let's stop and think about this. The classic notion of relying on the "heat of fermentation" of forages to relieve cold stress presumes that the rumen fermentation of forages generates more heat than the rumen fermentation of grains. Whoa — is there a disconnect here? If an animal's main source of body heat is the cellular metabolism of carbohydrates, and cold stress reduces rumen fermentation of fiber which reduces the heat of fermentation of forages, then why would we rely on this heat of fermentation to combat cold stress? In other words, why would we feed *more* hay to cold animals? Shouldn't we really be concerned about an animal's *body condition* — which reflects the amount of stored fat — and also about the flow of readily digestible carbohydrates into the system? Rather than feed more hay, shouldn't we feed grain?

Therefore, when animals are exposed to cold stress, instead of tossing them some extra bales of hay, we might consider *increasing the amount of grain in their diets*, at least during the period of cold stress. I'll point out that we're not trying to fatten these animals or risk a grain overload during this cold period; we're just giving them some extra readily-available carbohydrates to meet their increased need for heat production.

By the way, we're talking about *only a little extra* grain here; enough grain to provide a small boost of energy to cope with cold stress. The rest of our basic husbandry hasn't changed. We still feed hay or silage to our animals. Forage still furnishes most of the daily nutrients during this period. Forage keeps animals happily ruminating next to the feed bunk; it provides their bedding; and it gives us a warm sense of contentment on a cold winter night.

We must also be alert to particular situations of cold stress. Sometimes, our management techniques may put animals into cold stress by changing their LCT. For example, the practice of shearing ewes during late pregnancy. Although shearing pregnant ewes is a commonly recommended management technique, the shearing of winter-lambing ewes typically occurs during the winter. Slick-shearing can increase the LCT by nearly 30 degrees. We can modify these effects by shearing with cover combs or hand blades, which leave a thin layer of wool on the sheep. But these techniques only reduce the cold stress; they don't eliminate it. Ewes carrying twins must support the rapid growth of those fetuses during late pregnancy, and cold stress during this period — by claiming nutritional energy that would otherwise go to the fetuses — can really increase the risk of ketosis. In this situation, a little extra grain would go a long way.

It's a matter of perspective. A temperature of 35° during a Minnesota winter might seem warm, but 35° during a rainy Oregon winter would feel cold. Which one results in cold stress? It reminds me of a conversation I recently overheard at a livestock symposium:

> A professor asked the crowd, "*So, what is the most clever invention in the world?*"
> One shepherd quickly raised his hand and said, "*The thermos bottle.*"
> The professor was perplexed. "*Why?*" he responded, "*Of all the great advances in technology — penicillin, hot showers, Monday Night Football — why choose the lowly thermos bottle?*"
> "*Because a thermos bottle keeps hot liquids hot and cold liquids cold.*"
> "*What's so clever about that?*"
> "*How does it know?*"

First Published: December 1999

Author's Note: The "heat of fermentation" is a phrase I've heard since my first nutrition course. The concept dates back to the late 1800s — a little before my time — when our understanding of metabolism, fiber digestion, and rumen fermentation was rather rudimentary. We know more about livestock nutrition now. Even so, I still keep a thermos bottle of coffee in my pickup truck.

Pelleting in a Nutshell

There are many sides to pelleting. Some are straightforward, like prices and pellet size, but others are *not* straightforward, like the nutritional quality of the product. In fact, pellet nutrition is like a diamond — many tiny surfaces each reflecting a different facet of the topic. Some of these facets *increase* the feed quality of pellets, while others actually *decrease* it. Let's examine a few of them. Then we can better decide — to paraphrase a well-known, up-and-coming playwright — to pellet or not to pellet, because *that* is the question.

First, an important concept: pelleting is more than just forcing material through a die under heat and moisture and pressure. Pelleting also means *grinding* that material first, and grinding means smashing the fiber. Grinding demolishes a feedstuff's fiber *structure*. It's like smashing a wrecking ball into an old building. The building comes tumbling down and all the rooms are gone, but the original bricks and girders are still there, albeit in a somewhat different form. It's the same with feeds — grinding destroys the cell's three-dimensional structure, but all the fiber molecules are still there.

Why focus on *fiber*? Because pelleting and grinding affect the nutritional value of fiber more than any other portion of the feed. Fiber is the feed component that must first undergo fermentation in the rumen before it can be used by the animal. The non-fiber fractions of a feed, like sugar and fat, are already highly digestible, so grinding and pelleting really don't change them much. But because forages contain more fiber than grains, pelleting has a greater nutritional effect on forage-based pellets than on grain-based pellets.

Grinding smashes the fiber and exposes more of its surface area to the environment. When an animal consumes ground feed, this increased surface area gives rumen microbes easier access to the fiber particles, which speeds up the fermentation rate. This tends to *increase* the digestibility of the feed, although not always, because it depends on the type of fiber. Some types of fiber are not digestible at all, like lignin, and exposing those types of fiber to easier microbial access changes nothing.

But pelleting also *decreases* the digestibility of a feed because a pellet is, uh, a pellet. That is, it contains *everything* in the feedstuff that was forced through the pelleting die. A pellet contains *all* parts of a forage, including the low-value stems as well as the high-value leaves. When animals are fed loose hay in a feedbunk, they can usually sort through the forage and choose the best stuff. They can't do this with pellets. Pellets are an all-or-nothing feed, which means that animals are *forced* to eat poor-quality fibers which they would otherwise avoid. Feeding pelleted forages without including a supplement can effectively *lower* the nutritional value (TDN) of a diet. For some animals with high nutritional requirements, like lactating ewes or cows or goats, even a small decline in TDN value can be disastrous.

But pelleting also *increases* feed intake. Anyone who has ever fed pellets knows this. Animals eat more pelleted hay than loose hay, partly because smashed-up fiber doesn't take up as much room in the rumen, and partly because animals can wolf down whole mouthfuls of feed without much effort.

Higher intake means more pounds (or grams or molecules or whatever) of digestible nutrients moving through digestive tract, which potentially *increases* the nutritional status of the animal.

But higher intake also has its downside — it increases the rate of passage of feed through the digestive tract (after all, the extra feed has to go *somewhere*), which reduces the retention time of fiber in the rumen. Which gives the rumen microbes less time to ferment the fiber. Which can *decrease* the digestibility of the feed, because some of that fiber will leave the rumen undigested.

Physically, the pellet manufacturing process also generates some heat. This may or may not affect nutrient quality — depending on factors like the amount of heat, the moisture content of the feeds, the type of forage or grain, etc. Heat can cause some proteins and carbohydrates to combine into Maillard products (caramelization) which *reduces* the digestibility of those proteins. On the other hand, heat may *increase* the digestibility of some other proteins in the feed by increasing their bypass value, which is good if the animals need those extra amino acids. Moist heat may also slightly gelatinize some forms of starch in the feed, which can *increase* starch digestibility in the lower tract.

If that's not complicated enough, the forage species itself makes a big difference; grasses and legumes can react differently to pelleting because they contain different levels of fiber and types of fiber. Grasses generally contain more fiber than legumes. Grass fiber consists mostly of cellulose and hemicellulose, which are potentially fermentable in the rumen and thus potentially digestible by the animal. On the other hand, legumes like alfalfa and clover contain less fiber than grasses, but legume fiber contains a much higher proportion of lignin, which is virtually indigestible. This means that, on an equal weight basis, grass fiber generally has a higher *potential* digestibility than legume fiber. Therefore, by reducing the retention time in the rumen, the pelleting process may affect the nutritional value of grasses more than legumes. But . . . if those grasses are quite mature, they will contain higher levels of lignin, which is not digestible or affected by rumen retention time. And that's when things *really* get complicated.

The bottom line: the *net* effect of pelleting is the *sum* of all these factors (and perhaps a few other nutritional facets, like type of previous feed, availability of supplemental grain for fermentation, etc.).

Add into this equation some key economic issues like costs (what, pelleting is not free?) and wastage. Pellets almost always cost more than the equivalent loose feed, but pellets have little or no waste. And we must also consider our labor situation and our facilities for handling these feeds.

To pellet or not to pellet? That is the question. So, when you come across a simple answer to that question, you might want to reflect that some answers are, well, too simple.

First Published: October 1999

Author's Note: You would think that something so common as pelleting would be simple and easy. If this were only true. The reality is intricate and multi-layered, like so much else in nutrition. Although many of these factors are predictable, their weighted sum often is not. Each situation, in essence, must be evaluated on a case-by-case basis.

Four Guys and a Barn Fire

Watch out for the Maillard Gang this summer. Across the country these guys sneak into barns and burn them down. Official police documents may list the cause as "suspicious" or "spontaneous combustion" but the real culprit is this gang. There are four members: Amadori, Schiff, Strecker, and their leader, Maillard. They're involved in all sorts of shady deals — but they like to specialize in hay. Our hay.

I'm talking, of course, about wet hay and barn fires. There's no such thing as "spontaneous combustion." The cause is very definite and predictable — it's the notorious Maillard Reaction. Here's the story:

The Maillard Reaction is a complex chain of reactions that can occur between soluble sugars (carbohydrates) and proteins in environments like moist hay. The Maillard Reaction produces a brown, gooey substance that looks a lot like caramel. In fact, it *is* caramel. Two other names for the Maillard Reaction are "non-enzymatic browning" and "caramelization" — the same caramelization that, under the controlled cooking conditions on your stove, creates those chewy, glossy tan confections that bring back pleasant memories of summer beaches and county fairs.

Here's where the four guys come in. The *Maillard* Reaction begins when soluble sugars in moist hay combine with amino acids in that hay to form something called a glycosyl amine. This compound is rather unstable, and its atoms quickly do-si-do into an *Amadori* Rearrangement which then releases one water molecule to form a *Schiff* Base. This molecule then transforms itself in a reaction known as the *Strecker* Degradation, by combining with another amino acid and losing a molecule of carbon dioxide. The resulting compounds then polymerize with each other — a process similar to the making of nylon — to form something that looks like an amorphous, brown, yucky mess. This is the final product of the Maillard Reaction: the Maillard polymer.

To understand how the Maillard Reaction affects hay, we need to keep the following details in mind: Firstly, the Maillard Reaction gives off heat (an *exothermic* reaction). Secondly, the reaction goes faster in high temperatures. A rise of ten degrees will more than double its reaction speed. Thirdly, water acts as a *catalyst* to the Maillard Reaction, which means that the presence of water will increase the reaction rate. And finally, the Maillard polymer, which contains 11% nitrogen, is completely indigestible by sheep and cattle. Which means that all its nitrogen and carbohydrates are nutritionally lost to the animals.

Moist hay is prime territory for a runaway Maillard Reaction.

Let's stack some damp hay bales in a barn and see what happens (damp = 22–24% moisture or higher). Initially, some fungi and bacteria will grow on the moist leaves and stems. This is quite normal. They break down some of the carbohydrates and give off a little heat. Everyone knows that even good hay will heat slightly when first put into a barn (it "sweats"). But here's a kicker: water holds heat. The extra water in damp hay acts as a heat trap, preventing heat from escaping easily into the surrounding air. Because this damp hay contains so much moisture, heat begins to build up in the bales.

Within a couple of weeks, as the internal temperature rises above 130°–140°, the regular fungi and bacteria die off and some heat-resistant fungi begin to flourish. These fungi use up more carbohydrates and add to the heat load in the hay. As the temperature rises to more than 170°, other carbohydrates and proteins in the hay begin to combine on their own, without assistance from any living organisms or enzymes. This is the start of the non-enzymatic browning reaction — the Maillard Reaction — which begins to generate its own heat. The bale temperature soon climbs past 190°. At that point, even the heat-resistant fungi die off. Nothing can live in those bales. As the Maillard Reaction takes over, events now spiral out of control. The brown Maillard polymer begins appearing throughout the hay. Moisture accelerates the reaction, which begins to go faster and faster, generating even more heat. Temperatures climb rapidly. 300°. 400°. At 450–525°, the flash point of the hay is reached. The surrounding oxygen reacts with the hay. Fire. Explosive fire. Spontaneous combustion. Catastrophe.

If we can stop this process before it gets out of control — like dragging the warm bales from the barn — we'll still have nutritional problems with the hay, but we'll also still have a barn. The nitrogen locked up in that Maillard polymer is nutritionally unavailable to livestock. Which means that the hay's *available* protein is lower than its standard crude protein analysis. For example, if a heat-affected hay analyzes at 16% crude protein, and 25% of its nitrogen is tied up in Maillard polymer, the *available* protein value of that hay is only 12%. The hay will have a slightly yellow-brown tinge to it (hence the name "browning reaction"). A forage report on this hay would list the unavailable portion of the nitrogen as ADIN (acid detergent insoluble nitrogen) or ADI-CP (acid detergent insoluble crude protein). And we would consider this hay at 12% protein for the purposes of balancing rations.

So now what? Let's step back for a moment. The best way of dealing with the Maillard Reaction is to prevent it, so here are some basic rules of thumb: Fungi and bacteria don't survive when the moisture is less than 15%. Less than 15% moisture in any hay is the best of all worlds. (Well, we can hope for good hay-making weather, can't we?) In practice, however, to be safe, square bales should *always* contain less than 22% moisture. Less is better. Large round bales should *always* contain less than 18% moisture when stored in barns because large bales naturally retain internal heat longer than square bales.

And if you're in the market for buying someone else's hay, always look at the color. If you see a yellowing or browning, compared to normal hay, or in the worse case, a blackening, steer clear. Observe the hay under natural light because the bluish fluorescent light tends to mask subtle differences in color. Open up the bales. Dustiness means mold spores, which means that hay must have been wet enough to allow mold growth some time after it was cut.

Also look at the barn. Are there any holes in the roof that could have dripped water? And finally, is the barn still standing? If not, it's a sure sign that the Maillard Gang was there.

First Published: July 1998

Author's Note: Every now and then, as I drive down country back roads, I come across a loose pile of hay bales just outside a barn. Clearly someone had tossed them into a heap, in a hurry to move them out of the building. No additional explanation is necessary — that silent pile of hay says it all.

Old Hay, Good Hay?

There's a common belief that hay stored over a year old is no good — that it isn't worth very much.

Au contraire . . . Old hay can be a great value.

A few years ago I analyzed some old hay for a client. For nearly three years, he had stored a few leafy alfalfa bales in the back of his barn. That hay analyzed at 23.8% crude protein and 65% TDN, on a dry matter basis — roughly the same protein level as cull peas with an energy density almost as high as wheat bran. Not bad for old hay.

Therefore, can we really assume that old hay stored in a barn *automatically* loses nutritional value? I don't think so.

Consider what happens when we make hay. In essence, we take a three-dimensional snapshot of the growing forage and preserve it in a stable form for future use — kind of like a very large dried flower arrangement. So when can major nutritional losses occur with this hay? Well, most losses occur either while the hay is still in the field *before* it's hauled to the barn, or when the hay is fed to animals *after* the storage period.

In the field, when the green forage is freshly cut and laying in windrows, rain can cause mold growth and loss of nutrients. In very hot weather, excessive drying or raking will cause the nutritionally-valuable leaves to *shatter* (an impressive word meaning "to fall off the stems"). After the storage period, when the hay is fed, inefficiently-designed feeding facilities (or put less politely, *no* feeding facilities) are notorious for permitting up to 50% wastage of the hay. Also at feeding, wind can blow away the leaves, leaving a higher percentage of poorer quality stems. None of these losses, however, occur *during* the storage period when the hay is actually in the barn.

Now let's compare hay to its main alternative: silage. This is also a form of forage preservation, but instead of drying the forage, we pickle it. We take green or wilted forage directly from the field, stuff it into an airtight container, and let it rot. (Actually, another word for this process is *ferment*. Think of sauerkraut — made essentially with the same process.) Not much happens to the forage during the cutting and wilting process in the field, but major nutritional losses can occur *during the storage period.* Inside a silo, silage with too much moisture can lose runoff which contains soluble nutrients; a silage pH that is too high can result in major degradation of protein, not to mention an awful stench; air pockets in the silage will allow mold formation and nutrient loss; and when the silo bunker or bag is opened during the feeding process, the exposed forage will mold if the silage is not consumed fast enough. It's a dangerous world out there.

But can't hay also lose nutritional value during storage? Of course it can, but these losses are relatively easy to avoid by following two rules: (1) don't put wet hay into the barn, and (2) once you successfully put dry hay in the barn, keep it that way.

If stored hay contains more than, say, 15–18% water, mold can grow on it and use up some of the hay's available energy and protein. But mold growth also leaves telltale evidence — a few quadzillion mold spores, a musty odor, and possibly a yellowish or brownish color that indicates caramelization. This discoloration is due to a *Browning Reaction,* also called the *Maillard Reaction,* which is a chemical process that converts good-quality forage protein into indigestible caramel. This is the number listed in a forage analysis report as *heat-damaged protein.*

And here's a useful rule of thumb: don't cut a hole in the barn roof. Tightly-packed hay that becomes damp may mold excessively, heat very badly, and cause the barn to burn down. This is not a good way of controlling excess inventory.

But if stored hay is kept dry, not much bad happens to it. It just sits and sits and sits . . .

There is, however, one minor nutritional exception — Vitamin A. Green hay contains lots of Vitamin A when it is first put up, primarily as the compound *beta-carotene.* Although nutrients such as energy, protein, and minerals are quite stable in dry hay, the life expectancy of beta-carotene is relatively short. Over time, the carotenes are destroyed by oxidation, which is nearly impossible to prevent in dry hay. The bad news is that within a year, most of the hay's original Vitamin A activity has disappeared, even if the leaves are still green. But the good news is that Vitamin A is cheap. A farmer or rancher can go into the local feedstore and buy Vitamin A in a feed supplement, mineral mix, or as an injection.

If we can accept the premise that old hay in a dry barn can indeed retain its nutritional value, then we can begin to think like a bargain-hunting shopper at a garage sale. Old hay takes up storage space. A barn is nothing more than the farm equivalent of a warehouse, and farmers are like any other business people . . . they must move old inventory out of their warehouses to make room for the annual new models. Usually at a discounted price.

Therefore, all you have to do is locate good quality old hay. Shop around. When you find an attractive possibility, you can gauge the real value of that hay by doing a couple of things: Estimate the percentage of stems versus leaves — lots of stems indicate a mature forage. Look for the presence of legume flowers or grass seed — these suggest that the forage was harvested with a chainsaw. Also look for broadleaf weeds, which could reduce palatability.

Some other tips: Discoloration could indicate that the hay was put up damp. Shake some flakes or bales — do they release a lot of dust or have a musty odor? That "dust" is really a cloud of mold spores — more evidence that the hay had been put up wet. Look at the barn roof and walls — any signs of leakage or condensation? Look for rodent droppings, which in large amounts could affect palatability. And look for signs of barn cats. Cats can carry toxoplasmosis, which is an abortion disease spread through cat droppings.

So . . . let *everyone else* believe that old hay has a low market value. It's a buyer's market. Stock up on vitamin A injections or supplements, look around for a barn still filled with last year's crop, use your eyes and nose, and begin negotiating a price. Bring your checkbook. You may find a very good deal.

First Published: November 1995
Author's Note: When I talk about this topic in my workshops, audience questions come fast and furious. The belief that old hay automatically loses nutritional quality is so ingrained that many folks have a hard time believing otherwise.

A Sleuth in the Clover

Last summer I received a phone call from a rancher who had lost lambs under unusual circumstances. He was an experienced stockman who had run large commercial flocks for more than 25 years. As he began to talk, I had a feeling that his problem would be complex.

During that spring's lambing, his sheep had experienced a series of perplexing mineral problems. In one group of 90 ewe lambs which had lambed on a flat field of clover and grass, a number of lambs were born dead, some showing unmistakable goiters. A few weeks later in that same group of ewes, some young lambs died of white muscle disease. His veterinarians were confident about their diagnoses. In contrast, a different band of 100 ewe lambs on a nearby hill pasture lambed without any problems. The rancher had fed both groups the same loose commercial sheep trace mineral mixture that he had used in previous years. This mineral mix contained 84 ppm selenium (analyzed), a minimum of 70 ppm iodine, and no added copper. Earlier that winter, for the first time ever, the rancher had applied some fertilizer to the clover field. He asked me if the fertilizer could have caused these problems.

Multiple mineral deficiencies in animals consuming a balanced trace mineral mix? On a ranch operated by an experienced and knowledgeable rancher? I was right — this was not a simple problem.

This ranch is located in the Pacific Northwest, where the winters are cool and wet, and the summers are bone dry. Because winter temperatures remain above freezing, forages continue to grow slowly throughout those months. Then in the spring, forage growth explodes.

The two lambing areas were quite different. The flat lambing field contained ladino clover, annual ryegrass, foxtail, and some birdsfoot trefoil. The hill pasture contained alfalfa and brome. At least two weeks prior to lambing, the rancher applied fertilizer (0-21-21-7) to the flat field at a rate of 30 lb of actual phosphorus and 56 lb of actual potassium per acre. Enough rain fell after this application to wash off any residual fertilizer from the plants, which probably eliminated the risk of direct fertilizer consumption. In fact, that winter had been particularly wet, even by the soggy standards of the Pacific Northwest, with lots of rain and abnormally long periods of gloomy, overcast skies.

I asked about that pasture's soil characteristics and the forage growth. The flat field had a soil pH of 7.0. The soil analyzed at 7 ppm phosphorus, which is low, and 177 ppm potassium, which is generally adequate. The fertilizer contained no nitrogen because of the clover in that field. The rancher had observed, however, that the clover had not grown well during the spring — normally the ladino clover is a main forage at that time of year — and also that its regrowth after grazing was particularly poor. He also noted that the clover roots had not looked healthy: they were small with only a few nodules. In fact, when the rancher had applied some additional ammonium sulfate fertilizer to that field in the early summer, the clover looked greener and denser within 10 days.

Let's examine the obvious possibilities. Did the animals consume the trace mineral mixture? Yes. Are the rancher's observations reliable? Yes — this rancher was very experienced and professional. How much selenium did the mineral mixture *actually* contain? The feed tag listed 80–90 ppm, and the lab assay came back at 84 ppm. What about goiter from lack of iodine? The feed tag listed a minimum iodine level of 70 ppm, which is usually sufficient. Actually, I doubted that the mineral mixture lacked these nutrients because the group lambing on the hill pasture showed no problems with selenium or goiter. What about genetics? No . . . sheep genetics was probably not a factor because the same genetic lines had been on the ranch for years without these problems. Can fertilizer cause mineral deficiencies? Yes it can, but only rarely, and probably not for *both* problems at the same time. We've been looking at common factors, so now let's step outside the box and look at it from a different perspective — what factor was *not* common between these two pastures?

What about the forage? The hill pasture contained alfalfa and brome, but the flat pasture contained ladino clover which was not doing well. *That* caught my attention.

Ladino clover is a type of white clover (*Trifolium repens*). Also called *giant white clover* or *Italian clover*, it has larger leaves and a more upstanding profile than most varieties of white clover. As I thought about this forage, I remembered a passage from a book on toxic plants. White clover can sometimes contain very low levels of a compound called *linamarin*, which the rumen microbes enzymatically convert to cyanide. So what? Those lambs died of iodine and selenium deficiencies, not cyanide poisoning. But read on.

I suspected that during the spring the clover may have contained higher levels of cyanide than normal. Why? Because Linamarin levels can increase under certain stressful conditions — such as low light intensity, cool temperatures, and low levels of soil phosphorus. *All these factors* existed on this ranch during that winter and spring. The ladino clover was definitely under stress; its roots hardly contained any rhizobia-filled nodules, so it wasn't fixing much nitrogen. This was confirmed by its dramatic response to nitrogen fertilizer in early summer. Linamarin levels are also correlated to leaf size. Clover varieties with larger leaves tend to have higher background levels of linamarin. Ladino clover has large leaves.

Regarding selenium, the scientific literature contains reports showing that *cyanogenic glycosides* in plants can induce white muscle disease in lambs who have marginal levels of selenium. Linamarin is a cyanogenic glycoside. Although the trace mineral mixture contained 84 ppm selenium, which usually is enough, the continuous wet, muddy weather may have discouraged some ewes from consuming enough minerals, and their lambs may have been on the selenium edge, nutritionally-speaking. That's a stretch, of course, and it would almost have seemed too far-fetched if it weren't for the next item . . .

Cyanide has an intriguing metabolism in livestock. Animals are surprisingly adept at detoxifying low levels of it. (Apparently, cyanide is common enough in forages that animals have evolved a defense metabolism against small amounts of it. Larger amounts, of course, are deadly in a rather immediate and spectacular way.) Once cyanide is absorbed from the gut, it is transported into the liver where enzymes detoxify it by combining it with sulfur to form *thiocyanate*, which is then safely excreted in the urine. This thiocyanate, however, has an interesting characteristic — it is *goitrogenic*. In other words, continuously elevated thiocyanate levels in the blood prevent the thyroid gland from using the iodine in the blood, thus effectively causing an iodine deficiency, which shows itself as goiter. Bingo.

But what about the iodine in the trace mineral mixture? Two factors come into play here: amount and form. Although the guaranteed minimum *level* of 70 ppm wasn't bad, it was only slightly below the recommended level in minerals of 100–1,000 ppm. But the critical detail was the *form* of this iodine — calcium iodate. This mineral can easily be lost from the surface of mineral blocks. A loose mineral mixture contains far more surface area than a large mineral block. This ranch used a loose mineral mixture. Any loss of calcium iodate from the mineral mix translates to reduced iodine consumption by the ewes.

A picture slowly began to form — grim cloudy skies; a muddy pasture; the ladino clover under stress; the elevated levels of cyanide in that clover; the detoxification of that cyanide into thiocyanate; the goitrogenic effects of thiocyanate; a relatively low level of iodine in the mineral mix; the potentially higher loss rate of iodate from the mineral mix; the variable consumption of mineral by some ewes; the effects of cyanide on selenium status. A picture that precisely describes the problem faced by that rancher.

So, what to do about future lambings? Simple. Next year we plan to use a different source of iodine and put more iodine in the mineral mix. We'll replant the flat pasture with a legume species that does not contain cyanogenic glycosides. And as a contingency, we may also erect large sun lamps across the entire field if the weather becomes excessively cloudy.

First Published: September 2000
Author's Note: Sometimes I wish I had read more Sherlock Holmes. Pulling disparate items into a coherent solution is one of the more fascinating and enjoyable things we do as nutritionists. Reading, experience, and synthesis — that's what we're paid for.

SECTION 8

Intriguing Research

Spliced Alfalfa Genes

Last summer at the Animal Science meetings in Minneapolis I heard a report that really gave me a start. Some Australian scientists were trying to cross alfalfa with sunflowers.

Linda Tabe from CSIRO (*Commonwealth Scientific and Industrial Research Organization* — the Australian equivalent of the USDA Agricultural Research Service) described how her research group in Canberra was trying to increase wool growth by splicing sunflower genes into alfalfa. Huh? It didn't make sense to me either — at first — but after listening to the details, I began to think that this research is an incredibly innovative approach to managing the nutritional value of forages.

Australian farmers raise millions of fine-wool sheep on pasture. Every farmer would like to increase their profits by increasing fleece weight. This can be accomplished easily enough in a feedlot, where the diet can be carefully formulated, but how to do it in grazing sheep?

Wool fibers are composed of proteins, which are made from long strings of amino acids. Wool growth depends on a steady supply of amino acids from the diet. But unlike most other proteins, wool proteins also require a high level of specialized amino acids that contain sulfur (for the technically-minded: methionine and cysteine). Sulfur-containing amino acids are special because they can form interconnecting linkages between protein molecules, much like the scaffolding latticework at a construction site. These molecular linkages help make wool fiber very strong, but if a diet does not supply enough of these amino acids, wool grows slowly.

Forages, even high-protein species such as alfalfa and clover, have sulfur-containing amino acids that are not used efficiently by animals. Sheep raised on pasture have lighter fleeces than sheep that are fed supplements containing bypass protein. Alfalfa and clover may support excellent lamb growth, but they cannot support maximum wool growth.

So, how to get extra sulfur into sheep? Feeding bypass protein may *seem* sensible, but hand-feeding protein supplements to thousands of sheep does not. Daily supplementation of sheep on pasture is simply too impractical and/or too expensive for most farmers, especially on the extensive sheep stations in the Australian Outback. The problem of increasing fleece weight in grazing sheep, therefore, boils down to finding a practical way of delivering a daily dose of high-sulfur bypass proteins.

Linda Tabe's research group came up with a novel solution: search the plant kingdom for a plant that naturally contains a high-sulfur protein which remains stable in the rumen (i.e., a protein that is not degraded by rumen microbes), use modern gene-splicing technology to transfer the genetic code for this protein into a useful forage species, force those forage plants to manufacture the protein in their leaves, and put sheep onto the pasture to eat those leaves. Then just step back and watch the fleeces get heavier. Simple.

And maybe pigs will fly . . .

But these scientists indeed found such a protein. *Sunflower seeds* contain a storage protein called *Sunflower Albumin 8* that is particularly high in sulfur amino acids. Seed cells store this protein in bubble-like sacks called *vacuoles*, which are essentially convenient packages of stable high-sulfur proteins.

Although gene transfer is very common these days, gene transfer technology is still not quite as simple as, say, counting sheep. First, the Australian team had to transfer the sunflower genes to bacteria, which made many, many copies of these genes very quickly. Then, they successfully transferred the DNA from the bacteria into alfalfa and subterranean clover. So far, so good.

But sometimes science is a lot like ranching — sometimes an unexpected detail comes along to cause problems. This time, the detail was that pesky vacuole. In the world of biochemistry, vacuoles are not vacuoles are not vacuoles. Different cells use vacuoles in different ways: sunflower seeds use their vacuoles as stable storage areas; alfalfa leaves, on the other hand, use their vacuoles as garbage dispensers. Vacuoles in alfalfa leaves contain powerful enzymes that destroy any large molecules. (Sometimes I wish I had such enzymes to clean up my cluttered closets.) An alfalfa leaf vacuole is clearly not a good place to store anything.

But the Australian scientists countered with an alternate solution: by simply avoiding the vacuole. They modified the alfalfa genes so the cells would not transport the newly-manufactured proteins into the vacuoles. This allowed the sunflower proteins to remain in place at the cellular manufacturing site, where those proteins could accumulate safely.

The strategy worked. To their delight, the scientists observed that this genetic fix allowed alfalfa leaves to accumulate sunflower protein. Things worked out even better in subterranean clover. Subclover leaves manufactured higher levels of sunflower protein than alfalfa.

So now in Australia there are alfalfa plants and subterranean clover plants with multiple personalities. These plants grow like true forage legumes but they also manufacture rumen-stable sunflower proteins with high levels of sulfur. The prospects for improving fleece weight on pasture are looking better and better.

But we're not quite ready to put sheep onto those pastures yet. All these years of research have produced only a few *micrograms* of sunflower protein in alfalfa leaves. (1 microgram = 1 *millionth* of a gram. One gram = 0.035 ounce. That doesn't cover a lot of acres.) CSIRO scientists need to generate a lot more protein to do more research and answer some pivotal questions: Is this misplaced sunflower protein *truly* rumen-stable when sheep eat it? Does delivering these sulfur-containing proteins through alfalfa plants actually result in heavier fleeces? Do these genetic tricks unexpectedly reduce the growth potential of the alfalfa or subclover plants? And what about the economics of this thing; is it really worth it?

Linda Tabe's talk was the very last session on the last day of those Minneapolis meetings. As we milled around the room after her presentation, someone said, "The party's over. Last one out turn off the lights."

I beg to differ. This party has only just begun.

First Published: May 1995

Author's Note: This article was published in 1995. Transferring genes between species, of course, is no longer a cutting edge procedure. But changing the nutritional value of forages is still in its infancy. Since Linda Tabe first published her work, other researchers have successfully expressed these sunflower protein genes in many other plant species, including rice, lupins, white clover, and tall fescue. It's probably only a matter of time before one of these GMO (genetically modified organism) plants begins showing up commercially.

E. coli, Starch, and Feeding Grain

Jim Russell is a microbiologist at Cornell University. He and his colleagues recently published a scientific report about grain feeding and the survival of *Escherichia coli* (*E. coli*). By drawing on principles from microbiology, ruminant physiology, and animal feeding, they found something that could change how we formulate feedlot rations for cattle and sheep.

The bacteria *E. coli* usually live harmlessly in the rumen and colon (large intestine) of cattle and sheep. But one strain of *E. coli* — O157:H7 — is particularly nasty. If it gets into human intestines, 0157:H7 releases a highly virulent toxin that causes major damage to intestinal linings. Even just a few O157:H7 bacteria from contaminated foods (such as undercooked ground meat) will cause a week of cramps and bloody diarrhea called *hemorrhagic colitis*, and occasionally, this unpleasantness can escalate into serious kidney damage which can be fatal. It's this O157:H7 bacteria that is responsible for the *E. coli* outbreaks that periodically make the news.

But we don't get infected by O157:H7 simply because of contaminated meat. Human *E. coli* infections result from a complex series of events in the animal and in the human that include, possibly, feeding high levels of grain to the animal.

Some background: *E. coli* is actually beneficial in the digestive tracts of grazing animals. A thriving *E. coli* population in the colon prevents the growth of other, more harmful bacteria, and it also manufactures some B-vitamins. When animals are slaughtered for meat, packing plants go to great lengths to prevent the intestinal contents from contaminating carcasses. They separate out the intestines and scrupulously wash all carcasses. Still, it's almost impossible to prevent *all* contamination. Even near-microscopic particles may sometimes lodge in the meat. These particles may carry bacteria, including *E. coli*.

This situation would be only a minor problem because historically, most meat was sold as steaks and roasts. Since bacterial contamination occurs only on the *surface* of the meat, cooking sears the outside surface of the meat and sterilizes it, even if the chef prepares the meat so that the center is "rare." All this changed, however, with the huge growth of fast-food restaurants like McDonald's: suddenly the whole world wanted a hamburger. And billions and billions of hamburgers served meant that lots of meat was made into ground chuck instead of into steaks. But grinding meat mixes the surface of the meat with the interior, so anything on that surface — including bacteria — becomes mixed inside. And if the resulting ground meat is not cooked hot enough to sterilize the *inside* of the hamburger patty, those bacteria do not die.

Still, this is not enough to cause an *E. coli* infection. Humans are pretty robust creatures, and evolution has given us a few good defenses against food-borne bacteria. One key defense is our stomach, which is really a biological tub of hydrochloric acid with a pH of 2. That's a *strong* acid — strong enough to burn holes in blue jeans. Bacteria usually can't survive that acid, so our pre-digestive acid bath effectively acts as a sterilization chamber.

But an *E. coli* outbreak is unmistakable evidence that *E. coli* O157:H7 did indeed live through that acid bath, that something went wrong with our defenses. Jim Russell's group wanted to know why.

These researchers focused their attention on the effects of feedlot diets, which typically contain nearly 90% grain — corn, barley, wheat, milo (sorghum), etc. Grains are really compact packages of starch, and most starch is not very soluble in water. Cattle, as card-carrying ruminants, are really designed to eat lots of roughage that contains very little starch. But if cattle are fed a high-grain diet containing large amounts of starch, some of it will pass through the rumen unaffected by the rumen bacteria and flow downstream into the colon. This starch will support a high population of colon bacteria, which will happily ferment it, produce acids, and make the colon slightly acidic (to a pH of 5 or so).

Jim and his group sampled the colons from cattle fed either a hay diet or a 90% grain diet. Some of the results were entirely expected: colons from grain-fed cattle were slightly more acidic than colons from hay-fed cattle. But they pursued this investigation further. They made bacterial counts of those colons, extracted and purified the *E. coli* from them, and dumped those bacteria into a laboratory acid bath with a pH of 2. Their purpose was simple: they wanted to duplicate the acid stress that bacteria must face when they arrive in a human stomach. If *E. coli* didn't survive in a strong acid, they couldn't live long enough to cause damage in our lower tract.

Here's what these researchers found: Colons of grain-fed cattle contained a *thousand-fold more* coliform bacteria than the colons of forage-fed cattle and — most importantly — up to 10% of the *E. coli* from those grain-fed cattle survived the acid shock *while almost no E. coli from the forage-fed cattle survived the acid.*

These results were unexpected. The researchers concluded that *something* had changed in the *E. coli* bacteria in the grain-fed cattle. They speculated that the mild acid conditions in the colons allowed *E. coli* to adapt to acid — in some way *inducing* a biochemical mechanism to cope with the acid environment. When those same bacteria were then dumped into a bath of strong acid — simulating the conditions in a human stomach — that coping mechanism had already been activated and could help those bacteria survive the acid shock. In contrast, the *E. coli* from the forage-fed cattle did *not* survive the acid shock, probably because they didn't have a good way of coping with those acidic conditions.

The health problem for us humans is that only a few *E. coli* need to survive the stomach acid because it only takes a few O157:H7 bacteria to cause disease.

Does this mean the end of feeding grain in feedlots? Well, no. Jim's research only demonstrates that there are a lot of unanswered questions about the relationship between *E. coli* and high-grain rations.

But . . . at the very end of their report, and without providing many details, the researchers disclosed another interesting result: they could alter colon conditions by simply switching the diets. Changing the diet from a 90% grain ration to a hay ration dramatically reduced the *E. coli* population in the colon *after only five days.* Which opens up some tantalizing possibilities for solving the problem by merely changing the ration at the end of the feeding period.

No one wants an outbreak of O157:H7 *E. coli.* Five days of reduced growth is a very small price to pay for reducing the risk of infection. This may be research worth continuing.

First Published: December 1998

Author's Note: This research was published in 1998 and caused lots of controversy in the cattle-feeding world and also in the medical community. But things change slowly. As of this writing (2012), feedyards still feed high-grain rations right to the end of the feeding period. The impetus to alter those rations will most likely come from prohibitive grain prices rather than concerns for O157:H7. But in all fairness, since that original study, slaughterhouses have increased their scrutiny of carcass handling procedures and contamination risks.

BSE, Scrapie, & Bypass Protein

Let's discuss the relationship between bypass protein and scrapie.

First, some nutritional background: Bypass protein is protein that passes through the rumen without being degraded or fermented first by the rumen microbes. Why is this important? Because bypass protein can provide extra amino acids to high producing ruminants — like dairy cows during peak lactation or beef cattle gaining more than three pounds per day.

Bypass protein is sometimes necessary to support very high levels of milk or growth. All feed protein first passes through the rumen before it is attacked by digestive enzymes in the lower tract. Rumen microbes — bacteria and protozoa — ferment these proteins and reassemble most of the nitrogen into their own microbial compounds. (The nitrogen they cannot use is converted into ammonia which is excreted and lost to the system.) Most of the nitrogen used by the microbes becomes true protein in the microbial protoplasm and is fully digestible by cattle and sheep when it eventually flows into the small intestine. A small percentage of nitrogen, however, is entombed in the bacterial cell walls. This nitrogen is actually *indigestible* to ruminants because it's in a form that no lower tract enzymes can unlock. The bottom line is that rumen bugs convert a small percentage of feed protein into nitrogen compounds that are unavailable to the animal. This nitrogen loss can be considered as an overhead cost of rumen function.

The key concept here is protein solubility. Soluble proteins are fermented quickly by rumen microbes; insoluble proteins tend to pass through the rumen without being degraded. However, once insoluble proteins move into the true stomach and small intestine, they can be digested and absorbed just like any other proteins. Thus, insoluble proteins can provide amino acids more efficiently than soluble proteins because they are not subject to the rumen's overhead tax. In effect, these proteins "bypass" the rumen bugs; hence they are called *bypass proteins.*

Feedstuffs differ widely in their profile of proteins. For example, soybean meal and corn silage contain only small amounts of bypass proteins. On the other hand, three common feedstuffs are well-known for their high percentage of bypass proteins: cottonseed meal, roasted soybeans, and meat meal.

Meat meal comes from the rendering process, which is really an industrial cooking process. Rendering plants buy offal from packing houses and convert it into usable products. Rendering sterilizes the offal and separates the fat from meat and bone. The fat is sold as *tallow* (if it comes from cattle or sheep) or *lard* (if it comes from hogs). Renderers sell the non-fat product as *meat meal*, or if it contains higher levels of minerals, *meat and bone meal.*

Over the last twenty years, renderers have cultivated the dairy and feedlot industries as lucrative markets for these protein products. Modern dairy cows pumping out more than 100 pounds of milk each day, and feedlot cattle at high rates of gain, often require more protein than their rumens can supply. Simply feeding more protein will not work because of the overhead loss of rumen nitrogen. Bypass protein is a good nutritional solution, and rendered products economically meet this need.

Then in 1986 came a monkey wrench: England reported its first cases of *bovine spongiform encephalopathy (BSE)*, also known as "Mad Cow Disease." Cattle acted strangely, walking with an uncoordinated gate; docile dairy cows became erratic and hostile; and all the affected cattle died. Histology showed that the brains of these affected cows contained "spongy holes" and had plaque deposits of an aberrant protein called *prion*. Veterinarians noticed that BSE resembled scrapie in sheep. In fact, both diseases belong to a previously obscure family of unusual neural disorders called *spongiform encephalopathies.*

The official British veterinary report in 1988 concluded that no one knew how or why BSE suddenly appeared, but — here it comes — there *seemed* to be a possibility that BSE had *somehow* mutated from scrapie, and that the *most likely* source of infectious material was the rendered protein products from scrapie-infected sheep which had been used in cattle feeds. There was no experimental cause-and-effect evidence for this conclusion. Instead, scientists based their speculations on circumstantial evidence, because they had statistically eliminated other logical possibilities. (Their logic was indeed plausible. England has far more sheep than the United States, with a much higher occurrence of scrapie. Unlike in the U.S., British slaughter plants kill a relatively high percentage of scrapie-infected sheep, which would be reflected in the rendered products.)

This conclusion was a bombshell: the British government reacted quickly by outlawing the use of *all* ruminant-derived rendered products in ruminant feeds. No more recycling meat meal into cattle feeds.

Although the United States passed no BSE laws at that time, most of its renderers simply stopped accepting sheep. It was really a simple economic decision. BSE put American renderers on the spot: how to retain the confidence of their buyers while protecting their markets? No BSE had yet appeared in North America, and everyone, of course, wanted to keep it that way. So the renderers weighed the risk factors . . . as well as their customers' fears and the market impact.

The dairymen and feedlot owners demanded safety, or at least what they perceived as safety. No one anywhere knew for sure if scrapie *really* caused BSE, but on the other hand, no renderers wanted to lose their markets by frightening their customers. Feed companies could easily find alternative sources of bypass protein, and purveyors of cottonseed meal and roasted soybeans could *guarantee* that their products were risk-free. After all, who had ever heard of a scrapie-infected soybean plant? So in the United States, where sheep offal constitutes only a small fraction of the total tonnage of meat meal, the renderers decided that the sheep industry was economically expendable.

The current British panic about BSE has just brought all these issues to the surface again. This time, however, sheep are not suspect at all. Now the fear is that *human* cases of spongiform encephalopathies may be linked to *cattle*. The European Union has reacted dramatically: by banning all British beef. The USDA wants to institute a voluntary American ban on using rendered ruminant protein in any ruminant feeds, similar to the British rule of 1988. The American renderers, however, do not want to disturb their lucrative cattle-based market. Instead, they just want to ban sheep.

Go figure.

First Published: June 1996

Author's Note: I've included this article in the book because of its historical interest, and a chance to eat a little crow. Things change as we learn more. I originally wrote this article in 1996 during the British BSE panic and prior to the emergence of BSE in Canada and the U.S in 2003. In the years since 1996, researchers successfully demonstrated the linkage between recycled animal products and BSE. Governments worldwide — including the United States — instituted draconian regulations to prevent the recycling of rendered ruminant products back through ruminants, as well as extensive testing and monitoring programs. These actions have effectively reduced the BSE outbreaks and also the risks of occurrence. The primary objects of these activities have been cattle and humans. Sheep were just a small, and ultimately minor, part of the unfolding story.

SECTION 9

Reflections

Traveling Together

Our sheep industry is more than a series of topics on nutrition and breeding and health, and there are times to put away the technical formulas and think about the larger picture.

It's never been easy for me to step away from technology. As a student racing through graduate school, I was anxious, probing, driven — trying to rise above the course grades, trying to *see* the inner-workings of things. Weekends, weekdays, holidays. Late nights in the lab, weekends spent hunched over computer printouts and hand-scrawled notes on yellow pads. Then, after earning my degrees, working at various universities. More intensity, more weekend meetings, more travel. I built a portfolio: scientific papers, extension bulletins, magazine articles, presentations, software, videos.

I often crossed paths with people traveling down similar roads: fellow graduate students, faculty colleagues, farmers, ranchers, editors. Mostly good people, genuine and skilled. Usually we were too hurried to really get to know each other — activities were too hectic, a blur. But over the years, my concept of "the sheep industry" has gradually evolved — from a network of *events* and *organizations* into a fabric made up of *people*. It is the *people* who make these activities happen. Some of these people share my concerns and ideas and plans, and over time some have even become friends.

My friendships don't develop quickly. They often begin through technical issues — questions asked and answered, problems solved, programs conducted together. And perhaps by a shared sense of pride when we accomplish something good for our industry. After many events, over the years, some people seem to move closer to each other; building trust. They begin to rely on each other for support; they become part of a daily routine. Sometimes these relationships develop so smoothly and quietly that we hardly notice them.

Recently, one of my friends in the sheep industry became ill. The problem was serious, though repairable with surgery. But the illness was a rip in the fabric, a wavering of important things. The problem came on suddenly and unexpectedly, and when the illness was announced at a national meeting, the room became very quiet, a moment of silence, each person dwelling on his own thoughts. In the end, these meetings, these programs, these important functions are *not* just sterile events, they are gatherings of people who know each other, sometimes as friends.

But what I really want to talk about is that national meeting.

Hundreds of men and women from across the country gathered in a Denver hotel to decide on the future of an industry, our industry. They came from sheep farms and ranches, from businesses and government agencies. Some of these folks came from families who had been in the sheep business for generations, some for only a few years, but they all came to Denver for the love of an industry that was now in trouble.

The annual Board meeting of the national organization was a formal affair in a large hall. Most people wore sports jackets and dress suits. The tables had reservation signs with the names of the states. Scores of onlookers sat in rows of chairs behind the delegates, and the lights were dimmed to see the projector screen.

Two days of preliminary conferences had taken place prior to this meeting, informational gatherings where a new organizational plan was being presented and discussed. Now, during the formal annual meeting, it was time to vote.

But parliamentary procedure is a cumbersome affair. It's not a clean or efficient way to fine-tune a complex plan.

The organization president was from Utah, and he opened the session by outlining a set of ground rules; a reasonable method for working with an unwieldy issue. Someone else from Utah made the first proposal, a partial acceptance of the plan, but this caused immediate discussion which led to confusion. People were nervous — how could this formal body make the appropriate changes to something that needed delicate change? Someone who knew parliamentary procedure explained the options. The proposal was tabled. Another proposal was offered, which was also tabled. A North Carolina delegate proposed a sweeping alternative plan that took a different tack, one that satisfied many objections. The delegation from Wyoming supported this plan. The delegation from Pennsylvania did not. Each speaker went to the microphone to explain their positions, calmly, respectfully. More explanations came from other states. The North Carolina plan was special because it revised the bylaws and therefore needed a two-thirds majority to pass. A verbal vote was called for. Some yeas, some nays — it would require a written ballot. The lights came up, ballots were cast and counted — the proposal failed. More proposals, more amendments.

Then someone from California got up and gave an eloquent speech about the needs of the future, his vision. A new proposal was offered for a vote. It failed. This went on all afternoon, for hours, with motions, ballots, failed votes, amendments, short breaks, and tight little caucuses where people leaned in towards each other and tried to work out compromises.

Finally, near the end of this long afternoon, the original proposal was taken off the table and placed before the members for reconsideration. Amendments were added, some taken away. More discussion, more analyses, more options. Everyone was listening. Another vote. It passed.

Two roads diverged in a yellow wood,
And sorry I could not travel both
– Robert Frost

Our vote put us on a road, branching from a nameless junction that didn't have a sign. And now we're going down that road, with people we know and some we don't. And during this journey we'll use our skills and respect for each other and common sense and friendships, and we'll try to make it all work.

First Published: February 1999

Author's Note: The quote is from Robert Frost's poem "The Road Not Taken." Our industry is populated with strong-minded people with differing ideas. Some of those folks are friends; some are not. But Parliamentary Procedure was designed to allow people with different ideas to find ways to work out solutions. This meeting really happened in Denver in January 1999. We are still living with those solutions, and I still feel that we are traveling on this road together.

One of a Kind

Doug Hogue retired last fall. Some of you may not have known him. I did.

Doug has been a professor of animal nutrition and sheep production at Cornell University since the 1950's. Early in his career, he taught courses, conducted research, trained graduate students, and participated in the national affairs of the sheep industry. When the state Extension specialist retired in 1979 and the university administrators closed that position, Doug took over that outreach responsibility as well, and many New York farmers got to know him as the state sheep specialist.

I first met Doug in the summer of 1974. At the time, I was living in New York City and had decided to spend a week at Cornell to learn about its agricultural programs. Actually, I just wanted to learn how to drive a tractor. On the final afternoon of my Cornell visit, on a whim, I dropped by to see the person "who worked with sheep." Doug was in his office (I was blissfully unaware of rare occurrences). As soon as I entered the room, without any formal introductions, I asked why there weren't more sheep in New York State, since it contained so much grass? He looked up at this brash questioner and gave an answer. I asked more questions. I didn't know enough to shut up. After nearly two hours, he finally asked me who I was and why I was there. I told him that I was on campus to see about attending Cornell, but that I was worried about my chances of getting into the ag school because I hadn't grown up on a farm or attended an agricultural college as an undergraduate. As I chatted on about myself, he quietly measured me. He saw something that I was too unlettered to see. He paused, and then he advised me that entering Cornell was not impossible at all — "there are rules, and there are rules. We'll just go about it in a different way . . ."

The tractor lessons would wait. I did get into the Cornell ag school and earn a Master's Degree, and years later returned for a Doctorate. Both under Doug.

"We'll just go about it in a different way . . ." Think of the special type of person who says that.

Under Doug's guidance, I discovered a world of sheep and nutrition — a world where the glass is half full, not half empty. It's a world of opportunities and challenges and a confident person who can solve problems. A person with an original mind that often views things in, well, a different light . . .

Like the time Doug was an undergraduate at the University of California, Davis, and he took a physics course with its requisite homework assignments. Doug only worked on those questions for which he didn't know the answers. After all, why spend time doing exercises that he already knew?

Over the years at Cornell, Doug conducted research on topics that many producers use today, although not many people outside New York may associate his name with those works. For example, he did some of the early research on selenium requirements for sheep, when it was still news that selenium was a required mineral, not just a toxic one.

During his Cornell research in the 1980s, Doug was one of the first scientists to add fish meal to lamb diets — not as a protein replacement, but as a specialized source of bypass amino acids. At that time, everyone in the feed industry and most researchers confidently *knew* that fish meal was generally a bust as a protein supplement. It was too expensive for most least-cost rations; its strong flavor discouraged intake; and it quickly turned rancid in the feed trough. But Doug viewed fish meal as a potential source of certain amino acids that could pass through the rumen unscathed and augment the diets of fast-growing lambs. He conducted experiments where he added only a *small amount* of fish meal (3%) to carefully-designed 16% lamb diets — and produced higher growth rates and leaner lambs. Because of Doug's work, many commercial lamb rations in the eastern United States now contain a small amount of fish meal.

Doug also led breakthrough research on the early-weaning of orphan lambs. Two of his graduate students — Brian Magee and I — worked out many of the bugs of these procedures. The first orphan lambs I used in my experiments were Doug's own lambs — he hauled them from his farm in a gunny sack and a cardboard box. Today, thousands of sheep producers around the country can rely on cost-efficient procedures for weaning healthy lambs at less than three weeks of age — as a direct result of the knowledge gained from this research.

Over the last seventeen years, Doug and his university shepherd, Brian Magee, have quietly developed and refined the single most successful working strategy for accelerated lambing: the Cornell STAR System. Many people may not be aware of this, but back in the 1970's, the sheep industry decided that one of its most critical problems was the seasonality of lamb supply. Over the years, many researchers tried various approaches to this problem — breeds, hormones, light-treatments — with mixed success. But Doug and Brian approached the problem quite differently. They examined the physiology of sheep and built the entire production year around *it*. They divided the calendar year into five half-gestation periods of a ewe (each period 73 days) and put the rams in with the ewes at the beginning of every period. And they stacked the deck by using sheep breeds that tend to cycle naturally throughout the year. The STAR System takes three years for a single ewe to return to square one, which equates to five lambings in three years. Great minds are the first to cut through confusion and see things in a simple way. And at its most basic level, the STAR System is simple.

In the classroom, many undergraduate students, and more than a few graduate students, found Doug baffling or obtuse — he didn't always tell them precisely what to do. He informed, gave leads, provided resources, and then got out of the way. In other words, he cheerfully asked students to think. What a novel approach.

His expectations about people extended to his interactions with other professionals. At many academic seminars, I heard Doug ask seemingly innocuous questions. Sometimes, the speakers were job applicants who were giving a formal departmental seminar as part of their interview process. Doug's questions usually seemed simple enough, but they belied tough, ruthlessly logical probes. He gave the speaker an opportunity to demonstrate his knowledge about basic principles, or alternatively, enough rope to hang himself. There was no fanfare; no intimidation . . . just a soft-spoken question. But over the years, those questions helped build an animal science department of incomparable quality.

Innovative, strong-minded, enthusiastic, irreverent, professional — Doug Hogue is all of these. One can do worse than work with someone like Doug who can approach things "in a different way . . ."

First Published: May 1996

Author's Note: Read the dedication to this book. Our agricultural world would be a better place if more teachers and more researchers were like Doug Hogue.

Time, Land, and San Juan Hill

We live in a hectic time. As I write this, newspapers shout headlines about the latest convoluted politics in Washington, D.C. and about a monster hurricane swirling towards Florida with winds strong enough to blow over mountains. It could almost make one yearn for a different time. Now don't get me wrong — I'm not regaling the "good old days" — that fabled rosy age of blacksmiths under spreading chestnut trees and horse-drawn sleighs in the winter moonlight. That age also had no hot showers or antibiotics, so count me out.

But we in farming and ranching are fortunate. We may not be able to return to a different time but we can return to a different place, a quiet place — our land. Things move slowly on the land, almost without change, and the communities of folks who work the land also change slowly. Standing at the edge of a hay field, listening to the hum of summer insects and feeling the warm sun on my back — feels the same to me now as it did to a young man with a scythe newly returned from the Civil War in the summer of 1865. The land and our culture of the land are not hectic; they are quiet streams in a frantic world. So while we ruminate on this, here are two stories about time and the land.

Last summer I was driving a rental car in central Vermont, traveling between two places but with a few hours on my hands. As I drove north on Route 7, with the Green Mountains to my right and some rolling fields to my left, an old memory slowly formed in my mind. A long time ago — a couple of lifetimes it seems — I worked on a dairy farm in this area. Now I wondered if that place still existed.

I drove along Route 7 back and forth a couple of times, looking for a certain road to the west, trying to conjure up old scenes in my head. You know the feeling — you can't describe *exactly* what you are looking for, but you'll know it when you see it. A few miles north of Brandon, I found it — a narrow two-lane road dropping away from Route 7, lazily curving behind a cornfield. I turned the car. As I remembered it, the farm was only a mile or so down the road. I drove slowly, watching to the right, looking for the old barn and a white clapboard farm house with a porch.

Then I saw it, or at least most of it. There was the house, pretty much as I remembered, with a couple of shade trees in front, and just past it, a gravel driveway leading up to the barn. Except that the old barn was gone. Someone had torn it down and built a newer one just to the right — but the ancient concrete foundation was still there. In my mind, it looked about right.

So many years ago . . .

I had just returned from three years overseas in tropical Asia. I didn't have a job, but I wanted to work on the land, so I hitchhiked to Vermont. Everyone hitchhiked in those days, especially to Vermont. On a hot summer afternoon along Route 7, I got off my ride, hiked down the nearest country road, and walked into a farmyard where a family was stacking hay. A few words with the farmer and I became the hired man on his 50-cow dairy farm. For the next few weeks I worked there. I milked cows and fed calves and did whatever needed to be done on that small New England dairy farm that late summer. Although we had milking machines, we didn't have milk pipelines, so I hauled buckets of sloshing fresh milk from each cow to the milk tank, trying to keep the flies away. In the early mornings, I would stand at the east door of the barn and look out across the valley to the Green Mountains, watching the sun come up over those mountains. How long ago? Well, I distinctly remember listening with increasing horror to the news stories of black-hooded terrorists taking athletes hostage at the Summer Olympics in Munich. It was 1972.

Now I drove by the farm slowly. I could see some changes, and my eyes automatically measured the pastures. The fields were still rocky and unimproved, mostly overgrazed and showing green manure spots. The fence posts were sagging, but someone had placed step-in posts near them and strung a single electric wire to keep the cows in. I stopped the car and walked along the edge of the hay field. I felt the sun on my back and smelled the rich, dense scent of late summer grass. I took some photographs, but I didn't knock on the door of the house. I didn't know those people anymore, and I didn't want this day to intrude into my memories. After an hour, I drove away.

My second story occurred during the summer of 1975, when I spent a week working on an irrigated orchard in central Washington, a few miles west of Yakima. It was on a 30-acre family farm that grew pears, apples, and cherries. Beyond the farm boundaries, the surrounding land was hard and brown, a bleak range country, but inside, the orchard was a shady green island, the soil dark and moist from the life-giving irrigation water flowing down from the Cascade Mountains. My job was to thin the apples and pears by hand, to give the best fruit on each branch room to grow. It was wonderful to work under those shady trees, sculpting and molding each branch by removing the excess fruit. The hours moved slowly, and that was fine with me.

A middle-aged couple owned the farm, but the man's father was still on the place, and he enjoyed doing his share of the work under the trees. At 94 years old, he was still fit and strong. The two of us worked together that week, each taking separate trees but staying fairly close, each carefully selecting fruit and shaping the trees to get them just right. We often shared discussions and stories while we worked. I talked about my time overseas and my graduate studies in agriculture. His stories went farther back than that, about farming during the Depression and how the plains of central Washington looked before irrigation water came to the land. Our discussions always came back to the land.

The old gentleman was also a war veteran, and he sometimes talked about the war. His war. A time long past. His stories were sometimes humorous, sometimes not, but they were always interesting. I assumed, based on his age, that his war was World War I and that his stories were about Europe. But near the end of the week, as I carefully listened to his words, I suddenly realized that he was describing a geography that was not Europe. The city in his stories was not Paris. It was Havana. His war had been in Cuba . . . in 1898. He had been with the Rough Riders on San Juan Hill.

Nearly 30 years have passed since that orchard summer, and that gentleman is gone now. But the land remains. Orchard trees still cast inviting shadows, and I still like to stand by the edge of a hay field and feel the warm summer sun on my arms. I also like to think of others who have stood next to similar hay fields, at different times, in different places. Today's news broadcasts fade into the background. The land remains.

First Published: October 2004

Author's Note: These two stories are true. And when you get a chance, look at the haunting painting by Winslow Homer, *The Veteran in a New Field.*

Acronyms & Abbreviations

1,25D	1,25-dihydroxycholecalciferol
25D	25-hydroxycholecalciferol
AAFCO	Association of American Feed Control Officials
ADF	Acid Detergent Fiber
ADF-N	Acid Detergent Fiber Nitrogen
ADG	Average Daily Gain
ADI-CP	Acid Detergent Insoluble Crude Protein
ADIN	Acid Detergent Insoluble Nitrogen
ASI	American Sheep Industry Association
BSE	Bovine Spongiform Encephalopathy
CCT	Chronic Copper Toxicity
CF	Crude Fiber
CLA	Conjugated Linoleic Acid
CNCPS-S	Cornell Net Carbohydrate and Protein System for Sheep
CSIRO	Commonwealth Scientific and Industrial Research Organization
DAPA	Diaminopimelic Acid
DCAD	Dietary Cation-Anion Difference
DDG	Dried Distillers Grains
DDM	Digestible Dry Matter
DE	Digestible Energy
DIP	Degradable Intake Protein
DMI	Dry Matter Intake
DNA	Deoxyribonucleic Acid
DOM	Digestible Organic Matter
EDDI	Ethylenediamine Dihydroiodide
EE	Ether Extract
FDA	Food and Drug Administration

FNDF	Fermentable Neutral Detergent Fiber
GE	Gross Energy
GI	Gastrointestinal
GMO	Genetically Modified Oreganism
GRAS	Generally Recognized As Safe
GSH-px	Glutothione Peroxidase
HPLC	High-performance Liquid Chromatography
HTDN	Horse Total Digestible Nutrients
INDF	Indigestible Neutral Detergent Fiber
IU	International Units
iv	Intravenous
LCT	Lower Critical Temperature
LR	Lambing Rate
ME	Metabolizable Energy
MFL	Metabolic Fecal Loss
MJ	Megajoules
MP	Metabolizable Protein
ND	Neutral Detergent
NDF	Neutral Detergent Fiber
NE	Net Energy
NE_g	Net Energy for Gain
NE_l	Net Energy for Lactation
NE_m	Net Energy for Maintenance
NFC	Non-Fiber Carbohydrates
NFE	Nitrogen-Free Extract
NIH	National Institutes of Health
NIR	Near-Infrared Reflectance Spectroscopy (also NIRs)
NPN	Non-Protein Nitrogen
NRC	National Research Council
NSC	Non-Structural Carbohydrates
NSIP	National Sheep Improvement Program
ppm	Parts Per Million
RNA	Ribonucleic Acid
RXR	Retinoid-X Receptor
SID	Sheep Industry Development
STAR	is not an acronym. It refers to a 5-point star, from the calendar diagram of a specialized accelerated lambing system developed by Cornell University.
T2	Diiodotyrosine
T3	Triiodothryonine
T4	Tetraiodothryonine (Thyroxine)

TB	Tuberculosis
TDN	Total Digestible Nutrients
TM	Trace Mineral
TSH	Thyroid-Stimulating Hormone
UIP	Undegradable Intake Protein
USDA	United States Department of Agriculture
UV	Ultraviolet
VDR	Vitamin D Receptor
VFAs	Volatile Fatty Acids

Index

Editor's Note: Common nutrition and livestock terms including rumen, ruminant, feedstuff, forage, grass, corn, horse, cow, sheep, etc. have not been indexed, as they occur with great frequency throughout the book.

Made in the USA
Monee, IL
30 November 2024